HTC ThunderBolt™

FOR

DUMMIES®

by Dan Gookin

D1364349

WILEY

Wiley Publishing, Inc.

HTC ThunderBolt™ For Dummies®

Published by
Wiley Publishing, Inc.
111 River Street
Hoboken, NJ 07030-5774

www.wiley.com

Copyright © 2011 by Wiley Publishing, Inc., Indianapolis, Indiana

Published by Wiley Publishing, Inc., Indianapolis, Indiana

Published simultaneously in Canada

For general information on our other products and services, please contact our Customer Care Department within the U.S. at 877-762-2974, outside the U.S. at 317-572-3993, or fax 317-572-4002.

For technical support, please visit www.wiley.com/techsupport.

Wiley also publishes its books in a variety of electronic formats and by print-on-demand. Not all content that is available in standard print versions of this book may appear or be packaged in all book formats. If you have purchased a version of this book that did not include media that is referenced by or accom-panies a standard print version, you may request this media by visiting http://booksupport.wiley.com. For more information about Wiley products, visit us at www.wiley.com.

Library of Congress Control Number: 2011932041

ISBN 978-1-118-07601-9 (pbk); ISBN 978-1-118-13194-7 (ebk); ISBN 978-1-118-13196-1 (ebk); ISBN 978-1-118-13197-8 (ebk)

Manufactured in the United States of America

10 9 8 7 6 5 4 3 2 1

WILEY

About the Author

Dan Gookin has written more than 120 books about technology, many of them accurate. He is most famously known as the author of the original *For Dummies* book, *DOS For Dummies,* published in 1991. Additionally, Dan has achieved fame as one of the first computer radio talk show hosts, the editor of a computer magazine, a national technology spokesman, and an occasional actor on the community theater stage.

Dan still considers himself a writer and technology "guru" whose job it is to remind everyone that our electronics are not to be taken too seriously. His approach is light and humorous yet informative. He knows that modern gizmos can be complex and intimidating but necessary to help people become productive and successful. Dan mixes his vast knowledge of all things high-tech with a unique, dry sense of humor that keeps everyone informed — and awake.

Dan's most recent books are *Droid X For Dummies, Word 2010 For Dummies, PCs For Dummies,* Windows 7 Edition, and *Laptops For Dummies*, 4th Edition. He holds a degree in communications/visual arts from the University of California, San Diego. Dan dwells in North Idaho, where he enjoys woodworking, music, theater, riding his bicycle, being with his boys, and fighting local government corruption.

Publisher's Acknowledgments

We're proud of this book; please send us your comments at http://dummies.custhelp.com. For other comments, please contact our Customer Care Department within the U.S. at 877-762-2974, outside the U.S. at 317-572-3993, or fax 317-572-4002.

Some of the people who helped bring this book to market include the following:

Acquisitions, Editorial, and Media Development

Sr. Project Editor: Mark Enochs

Acquisitions Editor: Katie Mohr

Copy Editor: Rebecca Whitney

Technical Editors: Tapan Parikh, Keith Nowak

Editorial Manager: Leah Cameron

Editorial Assistant: Amanda Graham

Sr. Editorial Assistant: Cherie Case

Cover Photos: © iStockphoto.com / Richard Schmidt-Zuper

Cartoons: Rich Tennant (www.the5thwave.com)

Composition Services

Project Coordinator: Sheree Montgomery

Layout and Graphics: Samantha Cherolis, Tim Detrick, Corrie Socolovitch, Christin Swinford

Proofreaders: Lindsay Amones, Melissa Cossell, Melanie Hoffman, Lauren Mandelbaum

Indexer: Estalita Slivoskey

Publishing and Editorial for Technology Dummies

Richard Swadley, Vice President and Executive Group Publisher

Andy Cummings, Vice President and Publisher

Mary Bednarek, Executive Acquisitions Director

Mary C. Corder, Editorial Director

Publishing for Consumer Dummies

Kathy Nebenhaus, Vice President and Executive Publisher

Composition Services

Debbie Stailey, Director of Composition Services

Contents at a Glance

Table of Contents

Introduction

*I*f the phone is so smart, what does that make *you?* Really, you aren't a dummy, but using an advanced piece of technology such as the HTC ThunderBolt can make you *feel* like a dummy. Managing that much power, ability, and zing in such a teensy package can intimidate the most sophisticated of humans.

Though most people can get by, figure a few things out, or have their friends help with the little they know, you're about to take a different course: You have in your hands a book that will help you effortlessly leap the jargon-encrusted hurdles to understanding and using one of the latest, fastest, most technologically scary phones ever: the HTC ThunderBolt.

About This Book

Congratulations on reading this far into the Introduction. I've been writing technical books for years, and few people bother to read the book in order — Introduction first. It's as though those people would just walk into a cocktail party, ignore the hostess at the door, and head straight for the bar or perhaps that enticing snack table that holds those delightful little mints. Or hot wings. Yeah, I'd slam the hostess into a credenza to get at the hot wings.

Back to this book's Introduction, which I'm pleased that you're still reading: This book isn't intended to be read from cover to cover. The book is divided into parts, which are divided into chapters, which are minced up into sections. Each section is its own, self-contained unit representing something you can do with your phone, a task to perform, or an explanation of how something works. Sample sections include

- Typing on your phone
- Phoning someone you call often
- Forwarding a text message
- Sending a picture to Facebook
- Capturing video
- Listening to a tune
- Connecting to a Bluetooth peripheral
- Making the battery last longer

Still reading the Introduction? Good. Know that this book has nothing you need to memorize, and no spells to incant or crystals to clutch. Every section explains its topic as though it's the first or only thing you've read in the book. I make no assumptions, and everything is cross-referenced. The idea is to pick up the information you need and then close the book and get back to using your phone and getting on with your life.

How to Use This Book

As long you can read and know how to turn a page, you'll figure out how this book works in no time. Beyond that, it helps to understand that the ThunderBolt phone interacts with you, the human, by using a *touchscreen*. It's the glassy part of the phone as it faces you. There are also buttons on the phone, and along with the touchscreen, their operation is covered in Part I of this book.

You have various ways to touch the screen, which are explained and named in Chapter 3.

Text input on the ThunderBolt requires using an onscreen keyboard. See Chapter 4 for information on how to use it. Three keyboards are available, but you'll probably use the one that looks like a computer keyboard. I mention the other two, oddball keyboards in Chapter 4, but the rest of this book uses the standard *QWERTY* keyboard.

This book directs you to do things on your phone by following numbered steps. Each step involves a specific activity, such as touching something on the screen; for example:

3. Touch the Go button.

This step directs you to touch a graphical button, labeled Go, on the screen. You might also be told to do this:

3. Choose OK.

In this case, you touch the text or button labeled *OK*.

 Sometimes, a phone option can be turned off or on, as indicated by a gray box with a green check mark in it, as shown in the margin. By touching the box on the screen, you add or remove the green check mark. When the green check mark appears, the option is on; otherwise, it's off.

 When you're committing mass-deletion, an item can be flagged for removal as indicated by a gray box with a red X in it, as shown in the margin. As with the green check mark, touching the box adds or removes the red X.

To help you obtain software, or apps, for your phone, this book uses QR codes in the margin. A *QR code* is a type of barcode, though it's square, as shown in the margin, not striped, like a zebra.

To use the QR codes in the margin, you need a special app for your phone. See Chapter 26 for information about the Barcode Scanner app and how it's used. Chapter 18 covers the Android Market, which tells you how to obtain the Barcode Scanner app.

Foolish Assumptions

My most foolish assumption is that you're still reading this Introduction. I applaud you. Even my mom didn't make it this far. (She would have phoned.) Because you've read this far, I can make the assumption that you have a ThunderBolt 4G LTE phone manufactured by HTC and supported by Verizon in the United States.

Though the information here can be generically applied to any Android phone, it's specific to the ThunderBolt.

The ThunderBolt is a 4G LTE phone. The *4G* refers to the new, fast, fourth-generation data network. The *LTE* stands for Long Term Evolution, which is a technical topic that has to do with communications and I'm leaving the description at that.

As this book goes to press, 4G network coverage is limited, though by the end of 2012, you should be able to use your ThunderBolt at full 4G strength around the country — that is, unless those Mayan calendar predictions are true. Just in case, I'd wait until after December 21, 2012, before paying your monthly cell phone bill.

If you don't yet have 4G service in your area, the phone takes advantage of the 3G network. See Chapter 19 for more information on network data connection nonsense.

Another assumption I make is that you have a computer, either a desktop or laptop. The computer can be a PC or Windows computer or a Macintosh. Oh, I suppose it could also be a Linux computer. In any event, I refer to your computer as *your computer* throughout this book. When directions are specific to a PC or Mac, the book says so.

Programs that run on the ThunderBolt are *apps,* which is short for *applications.* A single program is an app.

Finally, this book doesn't assume that you have a Google account, but if you don't, you should get one. Information is provided in Chapter 2 about setting up a Google account — an extremely important part of using the ThunderBolt. Having a Google account opens up a slew of useful features, information, and programs that make using your ThunderBolt phone more productive.

How This Book Is Organized

Using a Ginsu knife (thank you, Shopping Channel), I was able to cleave this book into six parts. Each part describes a specific aspect of the ThunderBolt, how it's used, or what it can do.

Part I: A Phone, By Zeus!

This part of the book introduces the ThunderBolt. The chapters cover setup and orientation and provide an introduction to the phone and its features. Part I is the best place to start if you find the phone utterly confusing and prefer to crush it under your car.

Part II: The Phone Thing

It's called a *phone* because it does basic phone things. Duh. Part II covers making and receiving phone calls, plus other phone things, including forwarding calls, using voice mail, and maintaining a list of contacts. The contacts are used not only for making phone calls but also for sending and receiving email and doing other fun and interesting things.

Part III: Beyond the Basic Phone

You have ways of communicating beyond the basic phone call, and these basic ways are covered in this part of the book. You can send text messages, send and receive email, browse the web, and explore social networking opportunities. It's all covered in Part III.

Part IV: Superphone Duties

This part of the book answers the question "What else does this miraculous piece of technology do?" The ThunderBolt is capable of doing a slew of non-phone-like things. Activities covered in this part of the book include using the phone to find people and places in the real world, navigate your car, take pictures or video, use the phone as a digital music player, view videos, manage your schedule, and do other astounding things.

Part V: Nitty-Gritty Details

This part of the book contains chapters that cover a variety of topics, mostly details on using the phone, that just don't fit into the other parts. In this part, you can read about using wireless networking on the ThunderBolt, connecting the phone to a computer, sharing files, sharing the Internet connection, traveling with the phone, taking it overseas — plus, digesting some general maintenance and troubleshooting information.

Part VI: The Part of Tens

For Dummies books traditionally end with The Part of Tens, and this book is no different. I tried to make it different, but my brain started generating an incredible amount of heat as I tried to come up with 12 items per chapter instead of 10. So I stuck my head in a bowl of water and made some tea. With my head. It was great! But I ended up putting 10 items in each chapter of this part of the book, anyway.

Icons Used in This Book

This icon flags a helpful tip, suggestion, or shortcut.

This icon appears by information you shouldn't forget.

This icon marks something you should avoid doing.

This icon alerts you to information that's way too technical for a *For Dummies* book. You see, I can't help it: Occasionally, the inner nerd in me must be set free. This icon warns you about those occasions. Feel free to avoid reading any text near this icon.

Where to Go from Here

What you should do next, after you read this introduction, is start reading this book! You can use the table of contents to find a topic that piques your curiosity or use the index to hunt down something that's vexing you. If all else fails, avoid my advice and starting reading this book from cover to cover at Chapter 1. After all, you've read the entire Introduction. You're primed. You've earned it.

My email address is dgookin@wambooli.com. Yes, that's my real address. I reply to all email I get, and you get a quick reply if you keep your question short and specific to this book. Although I do enjoy saying Hi, I cannot answer technical support questions, resolve billing issues, or help you troubleshoot your phone. Thanks for understanding.

You can also visit my web page for more information or as a diversion: www.wambooli.com.

 Enjoy this book and your ThunderBolt phone!

Part I
A Phone, By Zeus!

In this part . . .

1 n exchange for their freedom, the Cyclopes gave the Greek sky god Zeus a gift: a thunderbolt. I'm sure that Zeus was thrilled. Not only did the thunderbolt become one of his primary symbols, but he also used it as a form of communications: Whenever a wayward Greek needed a reminder of his mortality, Zeus merely had to hurl a thunderbolt. It was a real attention-getter.

You probably don't want to lob your HTC ThunderBolt phone at anyone, no matter how obnoxious their hubris. Still, as far as a symbol of power, nothing beats a fancy, new 4G LTE phone like the ThunderBolt. Before you can wield your newfound authority, you need to liberate the phone from its box, get things set up, and understand how the thing works. Zeus probably went through the same orientation period with his own thunderbolts, but you have one thing he didn't: Part I of this book.

1

That New Phone Experience

In This Chapter

▶ Liberating your phone from its box

▶ Installing the SIM card and battery

▶ Charging the battery

▶ Identifying important things

▶ Reviewing optional accessories

▶ Keeping the phone in one place

I remember my first cell phone. I bought it in the early 1990s, and the gizmo cost me $600. The plan cost $45 per month plus $1.50 per minute of talk time. The cell signal didn't reach my house, so I could use it only in town — or if I wandered up to the roof and stood there during a call. Even so, I was thrilled to have a device unlike anything I'd ever owned, especially a gadget that let me do something wonderful, such as place a phone call from anywhere (well, anywhere it received a signal).

Your HTC ThunderBolt can be your first phone, your first cell phone, your first smartphone, and probably your first 4G LTE phone. It's an amazing gizmo, full of potential and loaded with things that will confound and confuse you. Before all that happens, you need to free the device from the confines of its box, give it a look-see, and complete other preliminary but important tasks.

Liberation

The first thing you do with a new electronic item is liberate it from the confines of its box or container. You repeat the same process for a new phone, though because most folks buy their phones at the Phone Store, odds are good that the people in the store have already freed your ThunderBolt from the confines of its HTC packaging.

Yes, it's akin to having your mom open your Christmas presents for you, but, as I was told at the Phone Store, it's "company policy." How devastating.

Whether the phone was manhandled by the salesperson in the Phone Store or you received the phone in another manner, this section covers removing the phone from its pretty box, setting up a few things, and getting the battery charged.

- Odds are good that the phone has already been configured to work on the Verizon cellular network. If not, you need to contact Verizon to get your ThunderBolt working. Someone there will be more than happy to help you set things up:

  www.verizonwireless.com

  (800) 922-0204

- The initial setup done at the Phone Store involves identifying the ThunderBolt with the cellular network. Basically, the phone's network ID is associated with the network and then magically mapped to your cell phone bill.

- Additional setup beyond configuring the phone for the cellular network involves setting up Google accounts for the ThunderBolt. This process might have been done with help from the people at the Phone Store, or you can do it on your own. This topic is covered in Chapter 2.

Unpacking the phone

Opening the box that the HTC ThunderBolt comes in can be a thrill. It's a pretty box, or at least the box my ThunderBolt came in had its charm. Opening the box is like opening a Christmas present or jewelry. It's fun.

A cling sheet of plastic on the front of the phone. It says, "Remove before use," which is printed on the cling sheet and not displayed on the phone's screen. The plastic is used for shipping purposes only, so feel free to peel it off your phone.

Along with the phone, you find some other items in the box:

- The battery, which may have already been installed into the phone by the cheerful Phone Store employee.

- A USB cable, which is used to connect the ThunderBolt to a computer or to a wall charger.

- A wall charger, or power adapter, which plugs into the USB cable.

- Pamphlets, warnings, and warranties that you can merrily avoid reading.

Free the USB cable and wall charger from their plastic wrapping because you need them in order to charge the ThunderBolt, covered later in this chapter.

▸ I recommend keeping the box for as long as you own your ThunderBolt. If you ever need to return the thing, or ship it anywhere, the original box is the ideal container. You can shove all those useless pamphlets and papers back into the box as well.

▸ One handy thing to have, which is missing in the box: headphones. See the section "Adding accessories," later in this chapter.

▸ If anything is missing or damaged, contact the folks who sold you the phone.

Inserting the SIM card

Before you can use your ThunderBolt, and before you install its battery, you must ensure that its SIM card is installed. The SIM card was most likely installed at the Phone Store. If so, great; you can skip this section. Otherwise, you need to install the SIM card, and you need to do so before you insert the battery (because the SIM card sits behind the battery inside the phone).

Obey these steps to install the SIM card into your ThunderBolt:

1. **If necessary, remove the phone's battery.**

   Refer to the next section, "Installing the battery," for removal instructions.

2. **If you haven't already, pop the SIM card out of its credit-card-like holder.**

   The SIM card ships as part of a larger piece of plastic, approximately the same size as a credit card. You need to pop out the SIM card before you can insert it into your ThunderBolt.

3. **Locate and slide out the SIM card carrier, found at the bottom of the battery compartment.**

   Use Figure 1-1 as your guide. The SIM card carrier works like a tiny, delicate, metal drawer.

4. **Place the SIM card into the carrier, as illustrated later, in Figure 1-6.**

   The notch in the corner of the SIM card sits in the lower left corner; it fits in only one way. The gold-plated connectors are face-down in the phone.

5. **Slide the SIM and its carrier back down into the bottom of the battery compartment.**

6. **Replace the battery and the phone's back cover.**

   See the next section for proper battery-replacement directions.

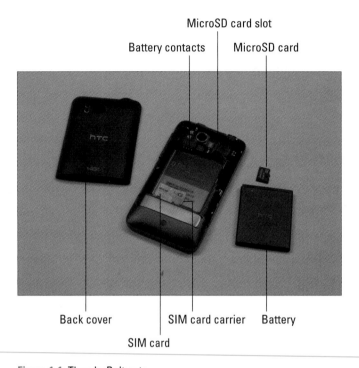

MicroSD card slot

Battery contacts   MicroSD card

Back cover   SIM card carrier   Battery

SIM card

Figure 1-1: ThunderBolt guts.

You'll probably never need to remove the SIM card from your ThunderBolt, but if you do, simply repeat these steps and lift out the SIM card in Step 4.

- See the nearby sidebar "SIM card nonsense" for more information on this SIM card nonsense.

- In a few instances, the ThunderBolt doesn't come with the SIM card preinstalled, such as when you're transferring service to a new phone or when you've ordered your phone from an outfit that doesn't supply SIM cards. Either way, you need to install the SIM card to use the ThunderBolt as a phone.

- The ThunderBolt requires a 4G, LTE SIM card compatible with Verizon's GSM network. You can't just use any old SIM card and expect the phone to work.

SIM card nonsense

The HTC ThunderBolt features a *SIM card,* which sets the phone's identity. The SIM, which stands for Subscriber Identity Module, contains a special serial number that's used by your cellular provider to help identify your phone and keep track of the calls you make. Additionally, the SIM can be used to store information, such as electronic messages and names and addresses, though you probably won't use this feature on your ThunderBolt.

A typical way to use a SIM is to replace a broken phone with a new one: You plug the SIM from the old phone into the new phone, and instantly the phone is recognized as your own. Of course, the two phones need to use similar cellular networks for the transplant operation to be successful. For the ThunderBolt, that means the other phone must be a GSM-based, 4G, LTE device.

On the ThunderBolt, it's the MicroSD card that's used as the phone's primary storage device, where you keep your music, photos, and other types of information as described throughout this book. But you still need a SIM card to make phone calls on the cellular network.

Installing the battery

As with the SIM card (see the preceding section), the ThunderBolt's battery may have already been installed by a cheerful Phone Store employee, depriving you of the opportunity of doing so yourself. If so, you can skip this section. If not, the battery comes loose inside the ThunderBolt's box and you need to install the battery inside the phone.

Follow these steps to install the battery into your phone:

1. **Ensure that the phone isn't connected to anything.**

   You should disconnect the phone from the USB cable, headsets, or anything else that might be attached.

2. **Flip the phone over so that the front (the glassy part) is facing away from you.**

3. **Remove the back cover.**

   The thumbnail notch at the top of the phone is illustrated in Figure 1-2: Stick your fingernail into the slot and pry off the phone's back cover. It takes some effort, but eventually the cover pops off, making a horrid crunchy sound as it's removed.

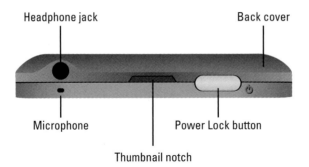

Figure 1-2: Removing the ThunderBolt's back cover.

4. **Set aside the back cover.**

   Marvel at the phone's guts, as shown in Figure 1-1.

5. **If you're installing a new battery, remove it from its plastic wrapping.**

6. **Orient the battery so that its metallic contacts are in the upper left corner as you're looking at the back of the phone.**

   There's a tiny arrow or triangle in the upper left part of the battery, which points at the contacts as the battery is inserted.

7. **Insert the battery top edge first, and then lower the bottom edge, like you're closing the lid on a tiny box.**

   Refer to Figure 1-3 for help in positioning and inserting the battery.

Figure 1-3: Inserting the battery.

8. **Replace the phone's back cover.**

   After you're oriented with the phone, press the back cover onto the phone. Keep pressing around the back cover's edges until you don't hear any crunching sounds.

After the battery is installed, the next step is to charge the battery, conveniently covered next.

Removing the battery works by repeating these steps, but with two changes: First, ensure that the phone is turned off before you remove the back cover. See Chapter 2 for directions on turning off the phone. Second, rather than insert the battery in Step 7, you yank it out: Insert a fingernail at the bottom left part of the battery, as illustrated in Figure 1-4. Lift the battery, like you're lifting the lid on a box, and then remove the battery.

Lift here.

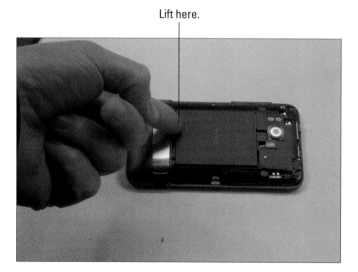

Figure 1-4: Removing the battery.

Installing or removing the battery isn't something you do often in your phone, but it's necessary in order to access the SIM or MicroSD card.

Charging the battery

The very first thing I recommend doing with your new phone is giving it a full charge. Assuming that the battery has been installed (see the preceding section), proceed with these steps:

1. **Attach the USB cable to the wall charger.**

   The cable plugs in only one way.

2. **Attach the USB cable to the ThunderBolt.**

   The cable plugs in only one way.

3. **Plug the wall charger into the wall.**

Upon success, you see an orange notification light appear in the phone's speaker mesh (at the top front part of the phone). It's the charging lamp, showing that the phone's battery is being rejuiced.

- Don't be alarmed if the orange notification light fails to appear.

- The notification light glows green when the ThunderBolt's battery has been completely charged.

- The ThunderBolt may have come fully charged from the factory, though I still recommend giving it an initial charge just in case, as well as to familiarize yourself with the process.

- The USB cable is used for both charging the phone and connecting it to a computer for sharing information, exchanging files, or using the ThunderBolt as a computer modem. See Chapter 20 for information on connecting your phone to the computer; see Chapter 19 for information on using the phone as a computer modem.

- You can also charge the phone by connecting it to a computer's USB port. As long as the computer is on, the phone charges.

- The battery charges more efficiently if you plug it into a wall, as opposed to charging it on a computer's USB port.

- You can charge the ThunderBolt using the power adapter in your car, if you have a car adapter charging accessory. See the section "Adding accessories," later in this chapter.

- The ThunderBolt's charging cable is a type of USB cable. The end that plugs into the phone is called a *micro-USB connector*. The other end is a standard *USB A connector*. Such a cable can be purchased at any computer or office supply store, or if you want to pay more money, you can get one at the Phone Store.

Examination and Orientation

Years ago, the telephone had a dial and a handset. The only ordeal you faced was determining which end of the handset to put to your ear and which end to put to your mouth. Things have changed in recent years, enough so that you need a more formal introduction to the various parts of your phone and what those parts are called.

Finding things on the phone

To peruse those things strange and wonderful on your phone, take a look at Figures 1-5 and 1-6. Every knob, hole, doodad, and smooth place has a name and purpose, as illustrated in the figures.

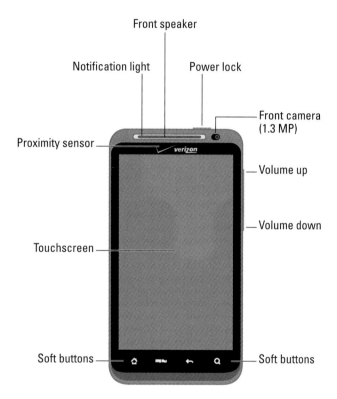

Front speaker

Notification light   Power lock

Front camera (1.3 MP)

Proximity sensor

Volume up

Volume down

Touchscreen

Soft buttons   Soft buttons

Figure 1-5: Your phone's face.

The items illustrated in Figures 1-5 and 1-6 are named using terms referenced throughout this book as well as in any ThunderBolt document you may have found with your phone or available online.

- ✔ The Power Lock button is found atop the phone, as shown in Figure 1-5 and 1-6. The button is used to turn the ThunderBolt on or off, as described in Chapter 2.

- ✔ The biggest part of the ThunderBolt is its touchscreen display, which occupies almost all the real estate on the front of the telephone. The touchscreen display is a touch-see gizmo. You look at it — and also touch it with your fingers.

Power Lock button

Noise-canceling microphones

Headphone jack

Dual LED flash

Main camera (8 MP)

Volume button

hTC

4G

Kickstand

USB connector

Figure 1-6: Your phone's rump.

- ✓ Festooning the area just beneath the touchscreen are four icons. These are the *soft buttons,* which are used to control the phone's software. You can read more about the soft buttons in Chapter 3.

- ✓ The back of the phone features a kickstand, which can be pulled out to prop up the phone for tabletop horizontal viewing.

- ✓ The kickstand covers a large, beefy speaker on the phone's rump. Deploying the kickstand uncovers the speaker, allowing for a hoard of people to be entertained by the ThunderBolt all at once.

- ✓ Use the Volume buttons on the right side of the phone to set the volume.

- ✓ Cameras are on both the front and rear of the phone. The front-facing camera can be used for video chat, though no app is preinstalled on the ThunderBolt to take advantage of video chat. Using either camera to shoot pictures or videos is covered in Chapter 14.

- ✓ Visit my website at www.wambooli.com/help/phone for information on video chat apps that may appear in the future, after this book goes to press.

> ✔ The front-facing camera is a 1.3 megapixel (MP) camera, which is good enough for video chat and pictures you send to the Internet. The rear camera is the main camera, which weighs in at 8MP. Its higher resolution allows for more-detailed images, but the phone's primitive lens doesn't take pictures that look as good as ones taken by a dedicated digital camera.

Using earphones

One item you need in order to use the ThunderBolt, something that didn't come in the box, is a pair of earphones. They help you use the phone in a hands-free manner, and they also come in handy for privately listening to music or other media.

The most common type of cell phone earphones are the *earbud* type: The buds set into your ears. The sharp, pointy end of the earphones (the part you don't want to stick into your ear) plugs into the top of the phone.

Between the earbuds and the pointy thing is a doodle. On that doodle, you find a teensy microphone hole. The doodle might also serve as a button, which you can use to answer the phone, hang up, call someone back, play or pause music, or do other potentially useful things.

> ✔ The earbuds are labeled L for your left ear and R for the right ear.
>
> ✔ The ThunderBolt can use any standard set of earphones; remember, though, that you want earphones with a microphone attached.
>
> ✔ The volume is set on the ThunderBolt by using the Volume buttons on the side of the phone.
>
> ✔ Some earphones feature extra buttons on the doodle, such as Pause, Play, Fast-Forward, and Volume. These buttons have no effect when used on the ThunderBolt.
>
> ✔ Chapter 16 covers playing music on your ThunderBolt phone.
>
> ✔ It's also possible to go wireless by using a Bluetooth headset or earphone. See Chapter 19 for information on connecting a Bluetooth headset to your phone.

> ✔ To avoid tangling the earphone cables, consider folding the wires when you put them away: Put the earbuds in one hand and the pointy thing in the other. Fold the wire in half, and then in half again, and then again. You can then put the earphones in your pocket or set them on a tabletop. By folding the wires, you avoid looping them into an impossible Gordian knot.

Removing and inserting the MicroSD card

The *MicroSD card* is the phone's removable storage device, like a media card in a camera or a USB thumb drive on a computer. You use the MicroSD card to store stuff on the phone, such as pictures, music, and contacts — just like on a computer.

The phone's MicroSD card is preinstalled at the factory; you don't have to insert the card when you first configure the ThunderBolt. The only time you need to remove the card is when you want to replace it with another card or when you need to remove the card to use it in another device.

To remove the MicroSD card, follow these steps:

1. **Turn off your phone.**

   Specific directions are offered in Chapter 2, but for now: Press and hold the Power Lock button (refer to Figure 1-5) and choose the Power Off command from the Device Options menu.

   If the phone isn't turned off, you can damage the media card when you remove it.

   To ensure that the phone is turned off, press and release the Power button quickly. The phone shouldn't come back to life. If it does, repeat Step 1.

2. **Remove the phone's back cover and then remove the battery.**

   Specific directions for removing the phone's back cover and battery are found in the section "Installing the battery," earlier in this chapter.

3. **Use your fingernail to drag out the MicroSD card, as illustrated in Figure 1-7.**

The MicroSD card is truly an itty-bitty thing, much smaller than your typical media card. I would advise not leaving the MicroSD card lying about.

To insert a MicroSD card into your phone, follow these steps:

1. **Ensure that the ThunderBolt is turned off.**

2. **Open the phone's back cover.**

3. **Remove the battery.**

4. **Orient the MicroSD card so that the printed side is facing up.**

5. **Gently insert the MicroSD card into its slot, using Figure 1-7 as your guide.**

6. **Reinsert the battery.**

7. **Put the back cover on the phone.**

MicroSD card

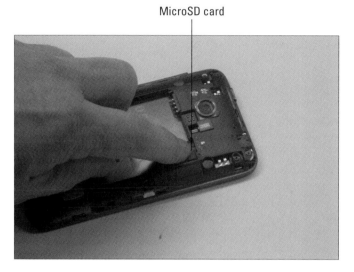

Figure 1-7: Accessing the MicroSD card.

After the MicroSD card is installed, you can turn on your phone. See Chapter 2 for details, though, basically, you just press and hold the Power button until the touchscreen comes to life.

- ✔ You can buy an SD card *adapter,* into which you can insert the MicroSD card. The SD card adapter can then be used in any computer or digital device that reads SD cards. Also, some USB adapters let you plug the MicroSD card into a thumb-drive-like device, which can be attached to any computer's USB port.

- ✔ SD stands for *Secure Digital.* It's but one of about a zillion different media card standards.

- ✔ MicroSD cards come in a smattering of capacities. The ThunderBolt ships with a 32GB MicroSD card, which is beefy, though you can buy a larger capacity if you and your credit card are willing.

- ✔ In addition to the MicroSD card, the ThunderBolt features internal storage, which is used for the programs you install as well as for the phone's operating system and other control programs. The internal storage isn't used for your personal information, media, and other items, which is why it's necessary to keep the MicroSD card inside your phone.

Adding accessories

It's hard to leave the Phone Store without first perusing the vast array of available phone accessories, including these standard items: earphones, vehicle chargers, genuine leatherette carrying cases, cables, and screen protectors. These items are all worthy of your attention, but two key accessories affect the phone's behavior: the Multimedia Desktop Charging Dock and the Window and Dash Vehicle Mount.

✔ No accessory is essential to using the ThunderBolt phone.

✔ Accessories also include various phone jackets and holsters.

✔ To find out more about earphones, see the section "Using earphones," earlier in this chapter.

✔ I recommend buying a screen protector. It clings to the touchscreen and truly helps keep it clean. Ensure that you get a screen protector designed for the HTC ThunderBolt and not for a lesser phone, especially one of those fruit-company phones.

The Multimedia Desktop Charging Dock

When nestled into the official HTC Multimedia Desktop Charging Dock, your phone displays a special multimedia menu, allowing you to play music or watch media while the phone charges.

The Multimedia Desktop Charging Dock also features an adapter that lets you charge a spare battery, if you have one of those for your ThunderBolt. (I suppose it's *another* accessory.)

The Window and Dash Vehicle Mount

There are quite a few car mounts and vehicle chargers you can get for your ThunderBolt. Any standard vehicle charger that features a micro-USB connector works with your phone. But of the lot of vehicle mounts, the Window and Dash Vehicle Mount from HTC is nifty.

In addition to allowing you to use the phone in both horizontal and vertical orientations, the Window and Dash Vehicle Mount is positioned in such a way that you can use the phone's main camera while you're driving. That way, you can record video of your trip (if the MicroSD card has enough storage). Make your own police chase videos!

Seriously, any car mount allows you to use your ThunderBolt while you drive. You can talk on the phone "hands free" or use the fancy navigation software to help you get where you're going. In fact, whenever the ThunderBolt is cradled in a car mount, the Navigation program, or *app,* runs automatically.

See Chapter 13 for more information on using your phone as your copilot.

A Home for the Phone

Your ThunderBolt phone ranks right up there with a lot of important things you use and don't want to lose, such as your car keys, glasses, wallet, and jet pack. As such, you should store and keep the phone in a handy, memorable location — even when you're carrying it around.

Toting your HTC ThunderBolt

The ThunderBolt is compact enough that it easily fits into your pocket or purse. The phone is designed so that you can carry it around without the risk of activating it or making one of those infamous "butt calls" — when you sit down and the phone in your pocket decides to dial up a friend.

The key thing to remember when carrying the ThunderBolt is not to forget that it's in your pants, purse, or coat. I consider myself fortunate that I've never sent my phone, in my pants pocket, through the wash. I have, however, tossed my coat on a couch and been mortified to watch the phone slide out and onto the floor.

- The ThunderBolt features a proximity sensor (refer to Figure 1-5), which disables the touchscreen while the phone is next to your face during a call or in your pocket or purse.

- Do not touch the Power Lock button when the phone is in your pocket. Doing so activates the touchscreen, making buttons such as End Call and Mute active.

- If you fear leaving your phone in your coat or purse, or sitting on the phone accidentally, consider buying one of those handsome carrying cases or belt clips. These accessories come in fine Naugahyde or leatherette.

- Also see Chapter 21 for information on using your ThunderBolt on the road.

- Rather than put your phone through the laundry, refer to Chapter 23 for proper phone-cleaning directions.

Storing the phone

I recommend finding a place for your phone when you're not taking it with you. Choose a location, such as on your desk by the computer, in the kitchen, on your nightstand, or in the same spot where you keep your car keys. The idea is to be consistent so that you can quickly find the phone when you need it.

✔ You can always find your phone by having someone call you and then hunting for the ring.

✔ A great place to store the phone when you're not using it is the Multimedia Desktop Charging Dock. See the section "The Multimedia Desktop Charging Dock," earlier in this chapter.

✔ I keep my ThunderBolt next to my computer when I'm not using it. This location has the bonus of my being able to attach the phone to the computer with the USB cable. I can charge the phone — and also synchronize my music, photos, videos, and other information while the phone is connected to the computer. See Chapter 20 for more synchronization information.

✔ Avoid storing the ThunderBolt in direct sunlight, because heat is bad for any electronic thingamabob.

✔ Don't put your phone on a coffee table or in a location where people stack magazines, the mail, or other items. You can too easily lose the phone in a pile of random stuff, or — worse — have someone do some cleaning and accidentally throw out the phone with the trash.

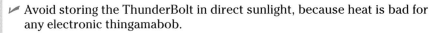

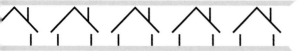

2

The On, Off, and Setup Chapter

In This Chapter

▷ Setting up and configuring your phone

▷ Unlocking the screen

▷ Waking up the ThunderBolt

▷ Getting a Google account

▷ Controlling Sleep mode

▷ Shutting down the ThunderBolt

*T*here should be nothing complex about turning on any device. Then again, the entire concept of something being turned on is a relatively new one for humanity. For example, Julius Caesar never had to learn how to turn on his chariot. To use a map, Genghis Khan merely unrolled it. And, try as you can, you won't find a single on–off switch on any of Leonardo da Vinci's inventions. Things are, however, different in the 21st century.

The ThunderBolt phone lacks an on–off switch. Nope, it has the Power Lock button. That's because the button does more than just turn the phone on or off. And, when you turn on the phone for the first time, you face some configuration and setup questions. All that nonsense is covered in this chapter.

Out of the Blue

The first question anyone asks when faced with a new technological gadget is "How do I turn it on?" This question is possibly followed at some desperate moment in the future by "How do you turn the silly thing off?" First things first: This section covers the various ways (yes, there are more than one) of turning on the ThunderBolt phone.

Turning on the ThunderBolt for the first time

The very, very first time your phone was turned on was most likely at the Phone Store. Or, if you bought the phone online, it might have been configured for you already. This basic setup is called *activation*. It involves connecting the ThunderBolt with the phone company's digital network as well as identifying the phone with your cellular bill.

After your phone's activation, the next thing you see when you turn on your phone for the first time is a prompt to configure your Google account. So, even if your phone is already activated, you still need to contend with the Google account setup. The following sections describe both procedures.

Activate your ThunderBolt phone

Follow these steps to turn on your ThunderBolt phone for the first time and activate your cellular service.

1. **Press the Power Lock button.**

   You may have to press it longer than you think; when you see the HTC logo on the screen, the phone is on.

2. **Unlock the phone; slide down the lock bar, as shown in Figure 2-1.**

Slide down to unlock phone.

Figure 2-1: The main Unlock screen.

What happens next, after you unlock the phone, depends on whether it has been activated. The remaining steps in this section assume that the phone hasn't been activated.

3. **Choose English.**

   Yes, if you're reading this in English, choose English as the language for your phone. *Si vous lisez ceci en français, choisir le français.*

4. **Touch the Activate button.**

   If you don't see the Activate button, the phone is already activated. Continue with Step 4 in the next section, which deals with configuring the Backup Assistant.

   The phone contacts your cellular provider to confirm that you have an active account.

5. **Obey the directions of your cellular provider.**

   The specifics of what happens next depend on your cellular provider. Often, you have to wait and listen to annoying and tinny music.

6. **Touch the Next button when you see the text *Device Is Activated.***

   The ThunderBolt may restart, depending on whether it was activated over the 4G LTE network.

If you have trouble activating the phone, contact your cellular provider. You need to read information from the phone's box, which has activation information printed on a label.

The lock bar shown earlier, in Figure 2-1, is one configured for use when the ThunderBolt ships. Other *skins* can be applied to the ThunderBolt, which change the lock bar's lock. See Chapter 22 for more information on personalizing your phone with skins.

Continue reading in the next section.

Set up your Google account

The next step in configuring your ThunderBolt is to synchronize the device with your Google account. As with activation, this step may have been performed at the Phone Store. If not, follow the directions in this section.

Your phone works in conjunction with a Google, or Gmail, account on the Internet. If you lack a Google account, see the section "Setting up a Google account," later in this chapter.

Follow these steps to configure your phone for use with your Google account:

1. **Turn on the phone, if it's not on already.**

   Press the Power Lock button until you see the HTC logo on the screen.

2. **Unlock the screen as described in the preceding section.**

   The Setup app, which configures the ThunderBolt for use with Google, runs automatically. If not, read the section "Attaching your Google account to the ThunderBolt," later in this chapter, to set up the phone for use with your Google account. (But then you need to come back to this step list, so keep your thumb on this page.)

3. **Choose English as your language and then touch the Next button.**

   You don't want to mess with the Backup Assistant, primarily because your Google account information is always automatically backed up. So:

4. **Touch the Skip button.**

   The Skip Backup Assistant warning window is displayed, mostly because the phone doesn't trust your willingness to be independent.

5. **Confirm your choice by touching the Skip Setup button.**

6. **Touch the Set Up Account button.**

   If you don't already have a Google account, see the section "Setting up a Google account," later in this chapter, and reread the paragraph by the Tip icon at the start of this section.

7. **Touch the Sign In button.**

8. **Touch the Username text box and type your Google account name.**

   Use the onscreen keyboard to type. Touch the buttons on the screen just as you would type using a real keyboard. Chapter 4 contains information on using the onscreen keyboard with your ThunderBolt.

9. **Touch the Password field and type your Google account password.**

   The password is hidden from view; as you type, black dots appear instead of the characters.

10. **Press the Back soft button to dismiss the onscreen keyboard.**

    The Back soft button is found below the touchscreen; its icon is shown in the margin.

11. **Touch the Sign In button.**

12. **Ensure that there's a check mark by the box labeled Back Up Data with My Google Account.**

    If no green check mark is in the box, touch the box to place a check mark there.

13. **Touch the Next button.**

    Read or ignore the text on the screen.

14. **Touch the Finish Setup button.**

    Ha! You thought you were done.

    The ThunderBolt enables you to set up additional accounts at this point in the Setup operation. You can add email and social networking accounts, though I describe how that's done elsewhere in this book. For now, you can merrily skip those steps so that you can start using your phone.

15. **Touch the Next button.**

16. **Ensure that check marks appear by each item on the next screen.**

    You see three items, each of which deals with how the phone knows where you are on Planet Earth:

    > *VZW Location Services:* Enable your cellular data provider (Verizon) to access location information, which aids in using mapping, location, and navigation software.

    > *Standalone GPS Services:* Allow location information to be accessed by the phone's programs as well as by the Internet.

    > *Google Location Services:* Allow Google to collect your location data anonymously.

    After touching the square to place a check mark, touch the OK or Agree button in the warning window that appears. It's okay to enable all the options, and, in fact, your phone works better when they're all turned on.

17. **Touch the Next button.**

    Now you're finally done, though you can touch the Keypad Tutorial button if you have a few extra, boring minutes that need filling. Otherwise:

18. **Touch the Finish button.**

     If you're returned to the All Apps screen, press the Home soft button to see the main Home screen and start using your ThunderBolt.

The good news is that you're done with setup. The better news is that you need to do this setup only once. From this point on, starting the ThunderBolt works as described in the next few sections.

- Information on your phone is synchronized with the information from your Google account on the Internet. Your contacts list, Gmail messages, calendar appointments, and other Googly things are all updated between the ThunderBolt and the Internet nearly instantaneously.

- If you've obtained any Android apps for another Android mobile device, they're installed on your ThunderBolt automatically.

- See Chapter 10 for information on adding more email accounts to your ThunderBolt. See Chapter 12 for information on setting up your social networking accounts. Chapter 18 covers Android apps.

- One of the first things you may notice is synchronized between your phone and Google is your Gmail inbox. See Chapter 10 for additional information on Gmail.

- When there's a lot of information in your Google account, such as hundreds of contacts or thousands of appointments on the calendar, it can take some time for everything to synchronize. Be patient.

- See the later sidebar "Behold the Android operating system" for more information about the Android operating system.

Turning on your phone

To turn on your ThunderBolt, press and hold the Power Lock button. After a few seconds, you see the HTC logo. The phone is starting.

Eventually, you see the main unlock screen, shown earlier, in Figure 2-1). Use your finger to slide down the Lock tab, as indicated in the figure. After the phone is unlocked, you can start using it — and unlike the first time you turned the thing on, you aren't prompted to go through the setup routine. (Well, unless you skipped setup the first time.)

- The Power Lock button is found on the top of the phone. Refer to Figure 1-5, in Chapter 1, for the specific location.

- You probably won't turn on your phone much in the future. Mostly, you're waking the gizmo from an electronic snooze. See the later section "Waking up the phone."

Behold the Android operating system

Just like your computer, your ThunderBolt has an *operating system.* It's the main program in charge of all the software (the other programs) inside your phone. Unlike your computer, however, Android is a *mobile device* operating system, designed primarily for use in cell phones but also in tablets such as the Motorola Xoom and the Samsung Galaxy Tab.

By using Android, the ThunderBolt has access to all the Android software — the *apps* or programs — available to Android phones and other mobile devices. You can read how to add these apps to your phone in Chapter 18.

Android is based on the Linux operating system, which is also a computer operating system, though it's much more stable and bug-free than Windows, so it's not as popular. Google owns, maintains, and develops Android today, which

is why your online Google information is synced with the ThunderBolt.

The Android mascot, shown nearby, often appears on Android apps or hardware. He has no official name, though most Android fans call him Andy.

Working the various lock screens

The unlock screen you see when you turn on (or wake up) the ThunderBolt isn't really a tough lock to pick. In fact, it's known as the None option in the Screen Unlock Settings window. If you've added more security, you might see any one of three additional lock screens.

The *pattern lock,* shown in Figure 2-2, requires that you trace your finger along a pattern that includes as many as nine dots on the screen. After you match the pattern, the phone is unlocked and you can start using it.

3. Continue tracing the pattern.

2. Drag your finger from dot to dot.

1. Start here.

Touch to make an emergency call.

Figure 2-2: The Pattern Lock screen.

The *PIN lock* is shown in Figure 2-3. It requires that you type a secret number to unlock the phone. Touch the OK button to accept your PIN, and use the Del button to back up and erase.

Finally, the *password lock* requires that you type a multicharacter password on the screen before the phone is unlocked. Use the keyboard, as shown in Figure 2-4, to type the password. Touch the green Enter button, illustrated in Figure 2-4, to accept the password and unlock the ThunderBolt.

Whether or not you see these various lock screens depends on how you've configured your phone's security. Specific directions for setting the locks, or removing them and returning to the standard screen lock, are found in Chapter 22.

Your code turn to dots for added security.

PIN code

Accept PIN and unlock. | Keypad

Back up and erase.

Figure 2-3: The PIN lock screen.

✔ You don't need to unlock the phone to answer it. See Chapter 5 for information on answering the phone.

✔ Even though the phone is locked, you can still make an emergency call: Touch the Emergency Call button, as illustrated earlier, in Figure 2-2.

✔ The ThunderBolt always has a lock screen. When a specific lock isn't chosen, the standard locking screen, as shown in Figure 2-1, is used.

✔ The pattern, PIN, or password lock may not show up if you've recently "slept" the phone. The lock appears only after a given timeout, which is 15 minutes unless you've set the value to something else. (See Chapter 22.)

✔ For additional information on working the onscreen keyboard, see Chapter 4.

Password

Onscreen keyboard   Back up and erase.

Accept password and unlock.

Figure 2-4: The password lock screen.

Waking up the phone

You'll probably leave your ThunderBolt on all the time. It was designed that way. The battery lasts quite a while, so when the phone is bored, or when you've ignored it for a while, it falls asleep. Well, it's technically in *Sleep mode*, or the special low-power, energy-saving state where the phone's touch-screen goes blank.

When the phone is sleeping, you wake it up by pressing the Power Lock button. Unlike turning on the phone, a quick press of the Power Lock button is all that's needed.

After waking the phone, you see the unlocking screen, as shown in Figure 2-1. If you've configured the phone for more security, you also see one of the unlocking screens, shown in Figures 2-2 through 2-4. Simply unlock the screen and you can start using the phone.

> ✔ The ThunderBolt continues to run while it's sleeping. Mail is received, as well as text messages. Music also continues to play while the phone sleeps and the display is off.

- ✔ Touching the touchscreen when it's off doesn't wake up the phone.

- ✔ Loud noises don't wake up the phone.

- ✔ The phone doesn't snore while it's sleeping.

- ✔ See the section "Putting the phone to sleep," later in this chapter, for information on manually snoozing the phone.

Account Setup and Configuration

If you've already configured your ThunderBolt to recognize and use your Google account, you're all done setting up and configuring the phone. That is, unless you've skipped that step or you need to create a Google account. Or, perhaps you need to set up the phone for use with your company's computer. Either way, directions are found in this section.

You can still pick up your email from Yahoo!, Hotmail, your ISP, or your business, as covered in Chapter 10. But to get the most from your phone, you need a Google account.

Setting up a Google account

If you don't already have a Google account, drop everything (but not this book) and follow these steps to obtain one:

1. **Open your computer's web browser program.**

   Yes, you should use your computer, not the phone, to complete these steps.

2. **Visit the main Google page at** www.google.com.

   Type **www.google.com** into the web browser's Address bar.

3. **Click the Sign In link.**

   Another page opens, where you can log in to your Google account. But wait — you don't have a Google account, so:

4. **Click the link to create a new account.**

   The link is typically found beneath the text boxes where you would log in to your Google account. As I write this chapter, the link is titled Create an Account Now.

5. **Continue heeding the directions until you've created your own Google account.**

Eventually, your account is set up and configured.

To try things out, log off from Google and then log back in. That way, you ensure that you've done everything properly and that you remember your password. (Your web browser may even prompt you to have it remember the password for you.)

I also recommend creating a bookmark for your account's Google page: The Ctrl+D or Command+D keyboard shortcuts are used to create a bookmark in just about any web browser.

Continue reading the next section for information on synchronizing your new Google account with the ThunderBolt phone.

Attaching your Google account to the ThunderBolt

Don't fret if you've failed to obey the prompts and neglected to sign in to your Google account when you're first configuring your phone. You can rerun the configuration setup at any time. Just follow these steps:

1. **Turn on or wake up the phone, if it's not already on and eagerly awaiting your next move.**

2. **Press the Menu soft button.**

   menu

   The Menu soft button is found beneath the touchscreen. Its icon reads "Menu," as shown in the margin.

3. **Choose the All Apps command.**

   The All Apps screen appears, which lists all programs, or *apps,* installed on your ThunderBolt.

4. **Scroll down the list until you find the Setup app.**

   Swipe or flick your finger across the touchscreen in an upward motion to scroll down the list. (See Chapter 3 for specifics on how to manipulate the touchscreen using your fingers.)

5. **Touch the Setup icon.**

   The Setup icon looks like a magician's hat, though the icon's look may change with updates to your phone's software.

6. **Work through the wizard as described earlier in this chapter.**

   Follow the steps listed in the earlier section "Turning on the ThunderBolt for the first time," in the subsection "Set up your Google account," starting at Step 3.

Your goal is to synchronize your Google account information on the Internet with the information on your phone. It all happens in a few seconds after you complete the Setup app.

Bye-Bye, Phone

Just as there are various ways to turn on the ThunderBolt, there's more than one way to turn the thing off — or not, as covered in this section.

Putting the phone to sleep

It's cinchy to sleep the ThunderBolt: Simply press the Power Lock button. The display goes dark; the phone is asleep.

- You can put the phone to sleep while you're making a call: Press and release the Power Lock button. The call stays connected, but the touchscreen display is turned off.

- Your phone will probably spend most of its time in Snooze mode.

- Sleep mode saves power because the display is turned off while the phone is asleep.

- The phone still receives email, text messages, and incoming phone calls, and even plays music while it's sleeping.

- Sleep mode doesn't turn off the phone.

- Any timers or alarms you set still activate when the phone is in Sleep mode. See Chapter 17 for information on setting timers and alarms.

- To wake up the phone, press and release the Power Lock button. See the section "Waking up the phone," earlier in this chapter.

Setting sleep options

You can manually snooze the ThunderBolt at any time by pressing the Power Lock button. In fact, it's called the Power *Lock* button because snoozing the phone is the same thing as locking it.

The phone automatically puts itself to sleep after a given period of inactivity, just like Grandpa Bob does during family gatherings. You can set the length of the inactivity period by following these steps:

menu

1. **From the Home screen, press the Menu soft button.**

2. **Choose Settings.**

3. **Choose Display.**

4. **Choose Screen Timeout.**

5. **Choose a timeout value from the list provided.**

   I prefer a value of 1 minute, which is the standard value.

6. **Press the Home soft button to return to the Home screen.**

The sleep timer begins after a period of inactivity; when you don't touch the screen, or press one of the soft buttons, the timer starts ticking. About ten seconds before the timeout value you set (Step 5), the touchscreen dims. Then it turns off and the phone goes to sleep. If you touch the screen before then, the timer is reset.

The phone doesn't sleep while you're watching a video, though it may sleep during some games — especially those puzzle games where I stare at the screen and scratch my head.

Turning off the ThunderBolt

To turn off your phone, heed these steps:

1. **Press and hold the Power Lock button.**

   You see the Power Options menu, as shown in Figure 2-5.

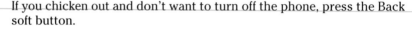

   If you chicken out and don't want to turn off the phone, press the Back soft button.

2. **Touch the Power Off item.**

   The ThunderBolt turns itself off.

The phone doesn't receive calls when it's turned off. The calls go instead to voice mail — as long as you've configured voice mail on the ThunderBolt. See Chapter 7.

Additionally, the phone doesn't sound alarms, collect email, or remind you of appointments while it's off.

Figure 2-5: The Power Options menu.

Basic Operations

In This Chapter

▶ Working the soft buttons

▶ Using the touchscreen

▶ Controlling the volume

▶ Getting around the Home screen

▶ Perusing notifications

▶ Running apps (programs)

▶ Reviewing recently used apps

*T*hose computers from old movies and TV shows were probably quite easy to operate. The reason is simple: They had too many buttons. There was a button on the console for everything. When the evil villain in a James Bond movie wants to secure all the exits to the underground lair, the henchman needs to flip only one switch. One switch! Such a high level of technology is truly amazing.

The ThunderBolt is quite a versatile device, despite its having only two buttons: Power Lock and Volume. Though the device lacks a specific button to seal the underground lair and prevent 007 from escaping, it's probably capable of doing that. The key is to understand the interface, which is why any evil overlord, henchman, or potential secret agent should read this chapter.

To Control a ThunderBolt

When someone says the word *telephone,* an image of the HTC ThunderBolt probably doesn't fly into your head. In fact, the ThunderBolt may be unlike any other phone you've ever owned. To master the device, familiarize yourself with some basic operations, as covered in this section.

Using the soft buttons

Four buttons are found below the touchscreen, each one identified by a unique icon. They're the *soft buttons,* which perform specific functions no matter what you're doing with your phone. Table 3-1 lists the soft buttons' names and functions.

Table 3-1		ThunderBolt Soft Buttons		
Button	*Name*	*Press Once*	*Press Twice*	*Press and Hold*
⌂	Home	Go to Home screen.	Go to the main Home screen or show Leap view.	Show recent applications.
menu	Menu	Display menu.	Dismiss the menu.	Do nothing.
←	Back	Go back, close, dismiss keyboard.	Do nothing.	Do nothing.
⌕	Search	Open the Search Anywhere app.	Open the Search Anywhere app.	Run the Voice Search app.

Not every button always performs the actions listed in Table 3-1. For example, if there's no menu to open, pressing the Menu button does nothing.

✔ Various sections throughout this book give examples of using the soft buttons. Their images appear in the book's margins where relevant.

✔ See the section "Accessing multiple Home screens," later in this chapter, for information on Leap view.

✔ Some online documentation refers to the soft buttons as *command keys.*

Working the touchscreen

The primary way you use your phone is to manipulate its touchscreen. As its name suggests, you do so by touching the glass with one or two fingers. I suppose if it were the *toescreen,* you'd manipulate it using your toes. But that's just not the case.

There are several techniques for working the ThunderBolt touchscreen:

Touch: The basic way to manipulate the touchscreen is to touch it. You touch an object, an icon, a control, a menu item, a doodad, and so on. The touch operation is similar to a mouse click on a computer. It may be referred to as a *tap* in the ThunderBolt documentation.

Double-tap: Touch the screen twice in the same location. The double-tap is used to zoom in on an image or a map, but it can also zoom out. Because of the double-tap's dual nature, I recommend instead using the spread or pinch method to zoom in or out.

Long-press: In a long-press, you touch part of the screen and keep your finger down. Depending on what you're doing, you may see a pop-up menu appear or the item you're long-pressing may get "picked up" and you can move it around. The long-press might also be referred to as the *press-and-hold* or *touch-and-hold* in some documentation.

Drag: The drag operation works similarly to dragging with a mouse on a computer. On a touchscreen, you press and hold and then, keeping your finger down, you move your finger — and an object — around on the touchscreen.

Swipe: To swipe, you start by touching one spot on the screen and then drag your finger in a certain direction: up, down, left, or right. The swipe is used to move the touchscreen content in the direction you swipe your finger. Swipes can be fast or slow. This technique is also called a *flick* or *slide*.

Pinch: A pinch involves two fingers, which start out separated and then are brought together. The effect is used to zoom out, to reduce the size of an image or see more of a map.

Spread: In the opposite of a pinch, you start out with your fingers together and then spread them. The spread is used to zoom in, to enlarge an image or see more detail on a map.

Rotate: A few apps enable you to rotate an image by touching with two fingers to the screen and twisting them around a center point. Think of turning a combination lock on a safe and you'll get the rotate operation.

You cannot manipulate the touchscreen while wearing gloves, unless they're specially designed for using the electronic touchscreen. You can buy this type of glove at the same place where Batman buys his wardrobe.

Setting the volume

The handiest way to set the ThunderBolt volume is to use the volume controls clinging to the upper right side of the phone: Press the top button to make the volume louder; press the bottom button to make the volume softer.

As you press the Volume button, a graphic appears on the touchscreen to illustrate the relative volume level, as shown in Figure 3-1.

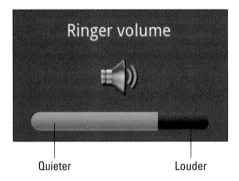

Quieter   Louder

Figure 3-1: Setting the volume.

When you press the Volume button and see the control, as shown in Figure 3-1, you hear a beep. It's your aural gauge of how loud the volume is set.

✔ When the volume is set almost to its lowest point, the phone is silenced. The Silence icon, shown in the margin, appears in the phone's status area.

See the later section "Examining the Home screen" for more information about the status area.

✔ Sliding down the volume level one notch below silent places the ThunderBolt in Vibration mode. The phone vibrates when it enters this mode.

When the phone is in Vibration mode, the Vibration mode status area icon appears, as shown in the margin.

✔ The Volume buttons work even when the phone is snoozing (in Sleep mode or when the touchscreen display is off). That means you don't need to awaken the phone if you're playing music and need to adjust the volume.

Refer to Chapter 22 for information on setting individual volume levels for the various apps, chimes, and activities on the ThunderBolt.

"Silence your phone!"

There are two social situations where you need to silence your phone. The first happens thoughtfully, at the start of a meeting, performance, or movie, where you're gently reminded to silence your phone and you diligently do so. The second happens in a panic as your phone disturbs the genial nature of a social gathering and you get flustered and rush to silence that damn ringtone.

To quickly silence the phone when it's ringing, simply touch the Volume button, up or down. You haven't dismissed the incoming call, so you can still deal with it: Answering or declining an incoming call is covered in Chapter 5.

To place the phone into Silent mode or Vibration mode, refer to the preceding section.

I lament that the ThunderBolt lacks a quick-and-dirty method for silencing the phone or placing it into Vibration mode. There's a widget you can get at the Android Market, called Silent Toggle Widget. It appears as an icon on the desktop when you touch once to silence the phone, again to place it in Vibration mode, and again to restore the sound.

See Chapter 18 for more information on obtaining Silent Toggle Widget; Chapter 22 covers installing widgets on your phone.

Touching the Volume button to silence an incoming call doesn't place the phone into Silent mode! The next call rings your ThunderBolt again, just as loud.

Changing orientation

The ThunderBolt features the *accelerometer* gizmo. It determines in which direction the phone is pointed, or whether you've reoriented the device from an upright to a horizontal position. That way, the information on the ThunderBolt's touchscreen display always appears upright, no matter how you hold it.

The easiest way to see how the ThunderBolt orients itself is to view a web page using the Internet app. Follow these steps:

1. **Touch the Internet app, found on the main Home screen.**

   The ThunderBolt launches its web browser. Unless you've messed with it already, the browser displays Google's mobile search page as the home page.

2. **Tilt the ThunderBolt to the left.**

   As shown in Figure 3-2, the image on the touchscreen rights itself, changing its orientation.

3. **Tilt the phone upright again, and then tilt it again to the right.**

4. **Press the Home soft button to exit the Internet app.**

Horizontal to the left   Vertical   Horizontal to the right

Figure 3-2: Vertical and horizontal orientations.

The phone displays an image horizontally on the left and right, but only vertically when you hold the phone with the Power Lock button on top; the ThunderBolt doesn't do upside-down orientation with the Internet app.

> ✔ This book shows the Home screen, as well as most other apps, in the vertical orientation. See the later section "The Old Home Screen" for more information about the Home screen.

> ✔ Not every app changes its orientation as you rotate the phone, but most apps do. Specifically, some games play in only one orientation.

> ✔ Wide-screen (horizontal orientation) is often the best way to view the web, as well as other apps that benefit from a wider display, such as Gmail and Mail.

> ✔ If the screen doesn't rotate, check to ensure that the Auto Rotate Screen option is set: From the Home screen, press the Menu soft button, choose Settings, and then choose Display. Place a check mark in the box by Auto Rotate Screen to ensure that apps change their orientation as you use the phone.

✔ A great application for demonstrating the ThunderBolt accelerometer is the game Labyrinth. It can be purchased at the Android Market, or you can download the free version, Labyrinth Lite. See Chapter 18 for more information on the Android Market.

The Old Home Screen

Operating your phone starts with a location called the Home screen. It's where the adventure starts.

The *Home screen* is the first thing you see after unlocking the phone, and the place you go to whenever you quit an app or end a call. Knowing how to work the Home screen is central to getting the most from your ThunderBolt.

Examining the Home screen

The main Home screen for the ThunderBolt is illustrated in Figure 3-3. It has several useful and curious things to notice:

Status bar: The tippy-top of the Home screen is a thin, informative strip that I call the *status bar.* It contains notification icons, status icons, plus the current time.

Notification icons: These icons come and go based on what's going on in your digital life. For example, new icons appear whenever you receive a new email message or have a pending appointment. The section "Checking notifications," later in this chapter, describes how to deal with notifications.

Status icons: These icons represent the phone's current condition, such as the type of network it's connected to, its signal strength, its battery status, and the current time.

Widgets: A widget is a teensy (or not so teensy) program that can display information, let you control the phone, access features, or do something purely amusing. You can read more about widgets in Chapter 22.

Application icons: The meat of the meal on the Home screen plate are the application icons. Touching an icon runs the application.

Notification icons

Status bar   Status icons   Widget

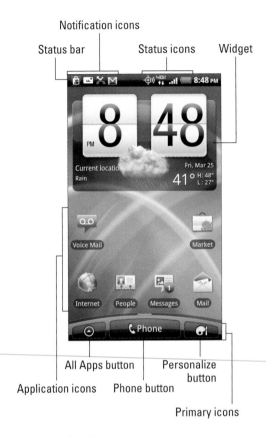

All Apps button   Personalize
button

Application icons   Phone button

Primary icons

Figure 3-3: The Home screen.

Primary icons: In Figure 3-3, you see the All Apps button, the Phone button, and the Personalize button. These buttons are the ones you use most often on the ThunderBolt, which is why they adorn the bottom of every Home screen panel.

It's important to recognize the names of the various parts of the Home screen because the terms are used throughout this book and whatever other scant ThunderBolt documentation exists. Directions for using the Home screen gizmos are found throughout this chapter.

✔ The Home screen doesn't do horizontal orientation.

✔ The Home screen is entirely customizable. Indeed, the ThunderBolt is one of the most personalized phones ever, which is why it has the Personalize button. You can add and remove icons from the Home screen, add widgets, make shortcuts, and even change the wallpaper images. See Chapter 22 for more information.

✔ Touching a part of the Home screen that doesn't feature an icon or a control doesn't do anything. That is, unless you're using the *live wall-paper* feature. In that case, touching the screen changes the wallpaper in some way, depending on the wallpaper that's selected. You can read more about live wallpaper in Chapter 22.

✔ The variety of notification and status icons is fairly broad. You see the icons referenced in appropriate sections throughout this book.

Accessing multiple Home screen panels

The ThunderBolt features a whole subdivision of Home screens. Out of the box, the phone is configured with seven Home screen panels, and the main Home screen is number 4, the center panel. Figure 3-4 illustrates the Home screen neighborhood.

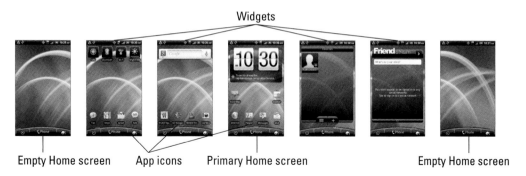

Figure 3-4: All the Home screens.

To switch to another Home screen panel, swipe your finger to the left or right. As you swipe, another Home screen panel pulls into view. The background image, or *wallpaper,* may even shift a tad as you switch between panels.

To see all the panels at one time, use Leap view: Press the Home soft button twice, or pinch your fingers when viewing the Home screen. You see a pre-view of all the panels, as shown in Figure 3-5. Touch a panel to see it at its full size.

Primary Home screen

Touch a thumbnail to hop
to that Home screen.

Figure 3-5: All the Home screens,
courtesy of Leap view.

You can rearrange the Home screen panels in Leap view. Any Home screen
can be swapped to a new position or made the primary Home screen. That
topic is covered in Chapter 22.

> ✓ No matter which part of the Home screen you're viewing, the status bar
> at the top of the touchscreen, and the primary buttons at the bottom of
> the screen, stay the same.

> ✓ To return to the Home screen at any time, press the Home soft button.
> Or, if you're viewing a panel, pressing the Home soft button returns you
> to the center Home screen panel.

The Home Screen Chore Chart

Just as there are basic household chores, there are some basic Home
screen operations. Your household chores might include tidying up, dusting,

vacuuming, scrubbing the toilet, or doing other things you routinely avoid. On the ThunderBolt, these operations are reviewing notifications, starting apps, and accessing widgets. I believe that you'll find them more enjoyable than doing housework.

Checking notifications

Notifications appear as icons at the top of the Home screen, as illustrated earlier, in Figure 3-3. To see the notifications themselves, you need to peel down the top part of the screen (see Figure 3-6).

Notification icons   Touch here.

Drag your finger down to
display the notifications.

Figure 3-6: Accessing notifications.

The operation works like this:

1. **Touch the status bar at the top of the Home screen.**

2. **Swipe your finger all the way down the touchscreen.**

   Think of this action as though you're controlling a roll-down blind; you grab the top part of the touchscreen and drag it downward all the way. The notifications appear in a list, shown in Figure 3-7.

Dismiss all notifications.

Recently opened apps   Swipe to see more.

Ongoing notifications (if any)
appear here.

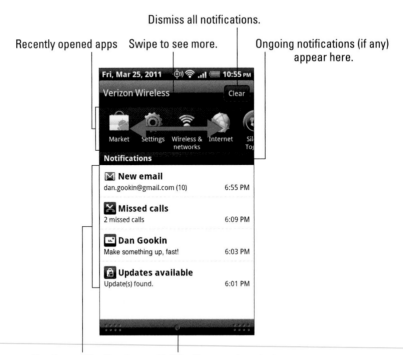

Touch a notification to
see more information
or to address an issue.

Notification panel control

Figure 3-7: The notification list.

Drag the notification list all the way to the bottom of the touchscreen to prevent it from rolling back up again. Use the Notification panel control to pull the list all the way down, as shown in Figure 3-7.

3. Touch a notification to see what's up.

Touching a notification icon displays the app that generated the notification. For example, touching a Gmail notification icon takes you to a new message or to your inbox.

If you choose not to touch a notification, you can "roll up" the notification list by sliding the Notification panel control back to the top of the touchscreen, or you can press the Back soft button.

⊯ If you don't deal with the notifications, they can really stack up!

⊯ The notification icons disappear after they've been chosen.

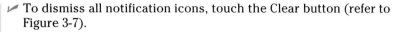

- To dismiss all notification icons, touch the Clear button (refer to Figure 3-7).

- When more notification icons are available than can appear on the status bar, you see the More Notifications icon, as shown in the margin. This icon always appears on the far left side of the status bar.

- The list of recently opened apps can be scrolled left or right with a touch of your finger. Touch an app to open the list.

- Touching the Clear button doesn't clear the list of recently opened apps. The list is reset when you restart the phone.

- You can also access the notifications list while viewing the Home screen by pressing the Menu soft button and choosing the Notifications command.

- Ongoing notifications show up when the phone is busy doing something, such as when the ThunderBolt is connected to a computer by using a USB cable. Ongoing notifications cannot be dismissed by touching the Clear button.

- Some apps, such as Facebook and the various Twitter apps, don't display notifications unless you're logged in. See Chapter 12.

- The ThunderBolt plays a sound, or *ringtone,* when a new notification floats in. You can choose which sound plays; see Chapter 22 for more information.

- See Chapter 17 for information on dismissing calendar reminders.

- Notification icons appear on the screen when the ThunderBolt is locked. Remember that you must unlock the phone before you can drag down the status bar to display notifications.

Starting an app

It's blissfully simple to run an application on the Home screen: Touch its icon. The application, or app, starts.

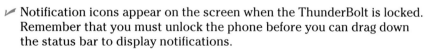

- Not all applications appear on the Home screen, but all of them appear when you display the All Apps screen. See the section "The All Apps Screen," later in this chapter.

- When an application closes or you quit that application, you return to the Home screen.

Working a widget

Like application icons, widgets appear on the Home screen. To use a widget, touch it. What happens next depends, of course, on the widget and what it does.

For example, touching the big clock widget on the main home page displays the Clock app or Weather app, depending on where you touch. Touching the Google widget displays the onscreen keyboard and lets you type, or dictate, something to search for on the Internet. Other widgets do similar things, with some widgets merely displaying information or letting you interact with them in an amusing manner.

Information about various widgets appears throughout this book. See Chapter 22 for information on controlling, adding, and removing widgets.

The All Apps Screen

The Home screen is a great place to keep a smattering of app icons, especially those you use most often. It doesn't, however, show all the apps on your phone. To find all your apps, you visit a location called the *All Apps screen.* Access it by touching the All Apps button at the bottom of the Home screen, as illustrated earlier, in Figure 3-3.

Viewing all your phone's apps

To start a program — an *app* — on your phone, heed these steps:

1. **Touch the All Apps button at the bottom of the Home screen.**

   The All Apps screen appears, as shown in Figure 3-8. The app icons are listed alphabetically, so the order changes as you install new apps on your phone.

2. **Scroll the list of app icons by swiping your finger up or down.**

   The list can get quite long, which is why it's advantageous to install app shortcut icons on the Home screen for the apps you run most often.

3. **Touch an icon to start its app.**

   Or, you can touch the All Apps button or the Home soft button to return to the Home screen.

The app that starts takes over the screen and does whatever marvelous thing that program does.

✔ See Chapter 22 for information on placing app shortcut icons on the Home screen.

✔ The terms *program, application,* and *app* all mean the same thing.

Applications

Touch to hide the   Swipe your finger up
All Apps screen.   or down to scroll.

Figure 3-8: The All Apps screen.

Accessing recently used apps

People commonly use the same apps over and over again. For example, you might check your stock portfolio or the scores of your favorite team, not to mention keep playing that annoying Jewels game.

 To easily access the apps you run over and over on your ThunderBolt, press and hold the Home soft button. When you do, the eight most recently accessed programs appear on the screen. To run one of these apps, touch its icon.

 To exit the recently used apps list, press the Back soft button.

You can press and hold the Home soft button in any application at any time to see the recently used apps list.

✔ You can also access recently used apps from the Notifications panel. See the section "Checking notifications," earlier in this chapter.

✔ For the programs you use all the time, consider creating shortcuts on the Home screen. Chapter 24 describes how to create shortcuts for apps, as well as shortcuts to people and shortcuts for instant messaging and all sorts of fun stuff.

4

ThunderBolt Typing

In This Chapter

▶ Using the onscreen keyboard

▶ Typing special characters

▶ Enjoying predictive text

▶ Editing text

▶ Selecting, copying, cutting, and pasting text

▶ Exploring alternative keyboards

▶ Dictating text with voice input

*H*istorically speaking, the typewriter had a short lifespan. The patent for the first typing gizmo was issued in 1714, yet truly functional machines didn't appear until the late 19th century. The typewriter itself ruled the office place from the early 20th century until 1990, when IBM stopped producing its ultimate typewriter, the Selectric. The typewriter died and the computer keyboard took its place.

The legacy of the typewriter is now a ghostly image on a flat piece of glass: To type on your ThunderBolt, you summon something called the *onscreen keyboard.* It looks like a typewriter, though it's flat and you touch the keys with your finger. Yes, not only has the old "hunt-and-peck" system returned, but using an onscreen keyboard is also the preferred way to work with text on your phone.

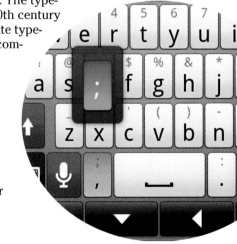

Everybody Was Touchscreen Typing

Your touch-typing kung fu skills are useless on a cell phone. Unless your hands are about the same size as a Barbie doll's, you have to type one finger at a time on the onscreen keyboard. Well, I suppose if you're truly deft, you can hold the phone in a manner that lets you type with both thumbs. I'm just not flexible enough for that action.

Exploring the onscreen keyboard

The onscreen keyboard is available anytime you have the opportunity to type text. The standard keyboard layout available on the ThunderBolt is the *QWERTY,* shown in Figure 4-1. It's the same layout you find on a computer keyboard, which is also known as QWERTY.

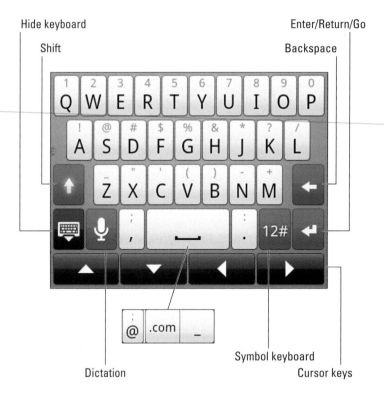

Figure 4-1: The standard onscreen keyboard.

When you're typing an email address, the semicolon-comma key changes to enable easy typing of the at-symbol (@). Also, the space key is bifurcated (divided) to provide space for the dot-com key, which, when pressed, generates the text *.com* on the screen.

To access symbols, press either the 12# or Symbol key. The onscreen keyboard changes to display common symbol characters, as shown in Figure 4-2.

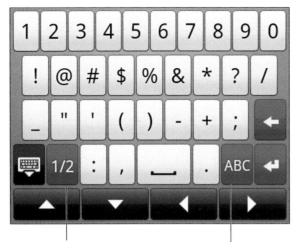

Touch to display the second symbol keyboard.

Switch to the alphabetic keyboard.

Figure 4-2: The first symbol keyboard.

The onscreen keyboard displays two palettes of symbols. Touching the 1/2 key, illustrated in Figure 4-2, displays the second palette, which is shown in Figure 4-3.

To return to the standard alphabetic keyboard, touch the ABC key (refer to Figures 4-2 and 4-3).

    ✔ The onscreen keyboard appears whenever you need to type text. If you don't see the keyboard, touch a text field or box where you type text and the onscreen keyboard pops up.

    ✔ You can use the Hide Keyboard button (refer to Figure 4-1) to temporarily dismiss the onscreen keyboard, making it vanish. The onscreen keyboard also can be hidden by pressing the Back soft button.

Touch to display the first symbol keyboard.

Switch to the alphabetic keyboard.

Figure 4-3: The second symbol keyboard.

- ✔ The cursor keys along the bottom of the keyboard are used for text editing, covered in the latter part of this chapter.

- ✔ You can also use the cursor keys to hop between links on a web page. See Chapter 11 for information on browsing the web on your phone.

- ✔ Touching any emoticon key shown in Figure 4-3 types all characters shown on the key. When you're sending a text message, these characters show up as special Android "cartoons." See Chapter 9 for details on text messaging on your ThunderBolt.

- ✔ QWERTY doesn't mean anything; it's simply the first six letters on the top row of the alphabetic part of the keyboard. Though on the planet Zavrox, the term *qwerty* describes a type of hand cramp.

Typing on your phone

Typing on the ThunderBolt's onscreen keyboard works as you expect: Touch a letter on the keyboard to produce that character.

To produce capital letters, use the Shift key, illustrated in Figure 4-1. Press and hold the Shift key to activate the shift lock (caps lock). In this mode, the Shift key's symbol changes, as shown in the margin. Press the Shift key again to turn off the shift lock.

The large U key at the bottom center of the onscreen keyboard is the Space key, which works like the spacebar on a computer typewriter.

The Enter key is used to end a line or paragraph of text, but it can also serve as the Go key when searching for or typing a web page address. The Enter key can move you from field to field when typing multiple tidbits of information.

- ✐ A blinking cursor on the touchscreen shows where new text appears, which is similar to how text input works on your computer.

- ✐ When you make a mistake, press the Backspace key to back up and erase.

- ✐ To hop from one text box to the next, use the cursor keys at the bottom of the onscreen keyboard.

- ✐ Above all, *type slowly* until you get used to the keyboard.

- ✐ See the later section "Text Editing" for more details on editing your text.

- ✐ When you type a password, the character you type appears briefly, but for security reasons it's then replaced by a black dot.

- ✐ When you see all capital letters on the alphabetic keyboard, the Shift key has been pressed.

- ✐ Some applications show the keyboard when the phone is in landscape orientation. If so, the keyboard shows the same keys, but offers more room for your stubby fingers to type.

- ✐ Not every application features a horizontal keyboard.

- ✐ When you tire of typing, you can always touch the Microphone button on the keyboard and enter Dictation mode. See the section "#### Typing: Use Dictation," later in this chapter.

Accessing special characters

Many keys on the alphabetic keyboard feature superscripted characters. To produce a superscripted character, press and hold the key. After a second, you see the superscripted character appear on the screen, as shown in Figure 4-4. Release your finger to type the character shown in green.

Pressing and holding a key, as illustrated in Figure 4-4, is often quicker than changing to the symbol keyboards, though the superscripted keys don't sport the variety of characters available on the symbol keyboards.

Refer to Figures 4-2 and 4-3 for the variety of symbols available when using the onscreen keyboard. Use the 12# and 1/2 or 2/2 keys, as illustrated in the figures, to access those variations of the onscreen keyboard.

To return to the standard, alphabetic keyboard (refer to Figure 4-1), touch the ABC key.

Press and hold. Superscripted characters

Character you typed

Figure 4-4: Typing a special character.

Enjoying predictive text

As you type text on the ThunderBolt, the onscreen keyboard helps you out (or not) by displaying suggestions. The suggestions may help you fix wayward typos or speed up typing by acting as shortcuts. It's the *predictive text* feature, and Figure 4-5 shows you how it works.

Suggestions

Word you're typing

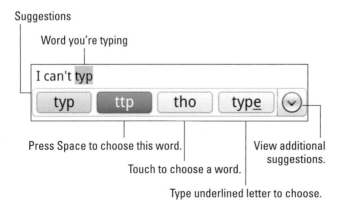

Press Space to choose this word.

Touch to choose a word.

View additional suggestions.

Type underlined letter to choose.

Figure 4-5: Predictive text in action.

As you type, word suggestions appear as buttons either above or below your text, as shown in Figure 4-5. Touch the Space key on the keyboard to instantly accept the word that's highlighted in green, or touch any word button to insert that word.

Additional word suggestions are shown by touching the Additional Suggestions button, illustrated in Figure 4-5.

You can add unrecognized words to the ThunderBolt dictionary, as described in the later section "Adding a word to the dictionary."

Disabling predictive text

Predictive text isn't to everyone's liking. Some people may find it annoying. If you're one of them, you can turn off the predictive text feature by following these steps:

menu

1. **From the Home screen, touch the Menu soft button.**

2. **Choose Settings.**

3. **Choose Language & Keyboard.**

4. **Choose the Touch Input command.**

5. **Choose the Text Input command.**

6. **In the QWERTY area (at the top of the screen), remove the green check mark by the Prediction item.**

7. **Touch the Home soft button to return to the Home screen.**

Likewise, if predictive text isn't enabled on your phone, activate it by repeating the steps in this section, but in Step 6, add the green check mark.

Adding a word to the dictionary

Occasionally, the predictive text feature encounters a word it doesn't know. When that happens, the word is highlighted in an orange button on the screen, as shown in Figure 4-6. You can continue typing, because the ThunderBolt doesn't care what you write, or you can choose to add the word to the dictionary by touching the Plus button, illustrated in the figure.

To ignore the word, just touch the Space key and keep typing.

If you choose to ignore the word, it continues to pop up as an orange button every time you type it. Therefore, I recommend adding unknown words to the ThunderBolt dictionary as you type them.

Also see Chapter 24 for information on viewing and editing the dictionary.

Press Space to accept the word.

Unknown, strange word   Add word to the dictionary.

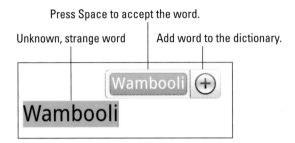

Figure 4-6: The ThunderBolt finds an unknown word.

Text Editing

Mark Twain was the first author to write a book using a typewriter. It was cutting-edge back then. I seriously doubt whether any author will write a book using the ThunderBolt. If you make the attempt, or for those more common situations where you simply need to edit text in an email message or on Facebook, you can use the text editing techniques covered in this section.

 A helpful app to use to test your text editing skills is AK Notepad. Obtain it from the Android Market, as described in Chapter 18.

Moving the cursor

The most basic of all text editing operations is moving the cursor. On your phone, the *cursor* is a blinking vertical line that marks the location where new text appears.

 To move the cursor, simply touch the spot in the text where you want the cursor to move. To help your precision, a cursor tab appears below the text, as shown in the margin. You can move the tab with your finger to move the cursor around in the text.

You can also finely position the cursor by using the arrow keys on the onscreen keyboard, shown earlier, in Figure 4-1.

After you move the cursor, you can continue to type, use the Backspace key to back up and erase, or paste in text copied from elsewhere. See the later section "Copying, cutting, and pasting" for more information.

 If you long-press the text, a pop-up magnifier window appears, which further helps you move the cursor. Releasing your finger displays a pop-up menu that can be used to select text, as described in the next section.

Selecting text

Selecting text on the ThunderBolt works just like selecting text in a word processor: You mark the start of a block, and then you select text to the end of the block. The chunk of text then appears highlighted on the screen.

Just to confuse you, you can select text in several ways, depending on what kind of text you're selecting and how precise you want to be. The following subsections uncover the details.

- ✔ To cancel text selection, or undo any selected block, press the Back soft button.

- ✔ Selecting text is only half the job. The other half is to copy or cut the text, which is covered in the later section "Copying, cutting, and pasting."

- ✔ You can delete a selected block of text by touching the Backspace key on the onscreen keyboard.

- ✔ Any text you type replaces the selected block of text.

- ✔ Seeing the onscreen keyboard is a good indication that you can edit and select text.

Select text with a double-tap

The quickest way to select a single word is to double-tap it using your finger. The text is selected, as shown in Figure 4-7. At that point, you can extend the selection by dragging the starting and ending markers for the block of text or use the Copy, Cut, or Paste menu items to manipulate the text, as described later in this chapter.

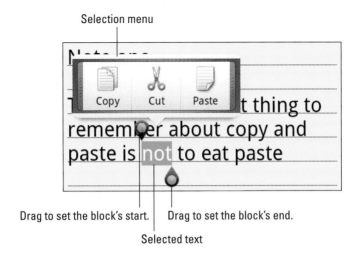

Figure 4-7: Selecting text.

Select text with the magnifier

When you long-press text, you see the magnifier window, which helps you locate the cursor. Releasing your finger displays a menu with these three items:

Select Text: Choose this item to enter Text Selection mode. A word is selected and you see a menu similar to the one shown in Figure 4-7. You can extend the text selection or choose a menu item to copy, cut, or paste text.

Select All: Choose this item to select all text in the document, on the screen, or in the text box. The menu options available after selecting all the text are Copy, Cut, and Paste.

Paste: Choose this item to replace the selected box with whatever text has been previously copied or cut. See the section "Copying, cutting, and pasting," later in this chapter, for details.

Text selection on a web page

You can select text from a web page, even though the text isn't "editable" and the onscreen keyboard appear doesn't appear. Because the text isn't editable (or even edible), you're limited in what you can do with it.

To begin selecting, long-press the web page. A single word is selected, similar to the one shown in Figure 4-7. You can extend the selection by dragging the start and end markers through the text.

Three commands are available for the selected block of text on a web page:

Copy: Choose this command to copy the text to the clipboard. From there, it can be pasted into any location where text is typed in an app on your phone. See the next section.

Quick Lookup: Choose this command to run the Quick Lookup app, shown in Figure 4-8. You can use this app to use the text to search Google, Wikipedia, or YouTube, or you can translate the text or look it up in an online dictionary.

Share Via: Choose this command to send the text elsewhere by using the Share Via menu. You can send the text as part of an email message or a text message, to Facebook or Twitter, or to other apps installed on your ThunderBolt, all depending on what shows up on the Share Via menu.

Choosing the Share Via item is an easier way to copy and paste text from a web page into an email message or another app, such as AK Notepad, mentioned earlier in this chapter.

- ✓ Alas, there's no simple way to select all text on a web page at one time.

- ✓ Sharing menus are popular on the ThunderBolt. You can find Share and Share Via commands in various apps, where you can use the menu to share information on your phone. Refer to Chapter 15 for information on the Gallery app, where I illustrate how to work several sharing options.

✔ You can only copy text from a web page. Obviously, you cannot cut text. See Chapter 11 for information on saving pictures from a web page.

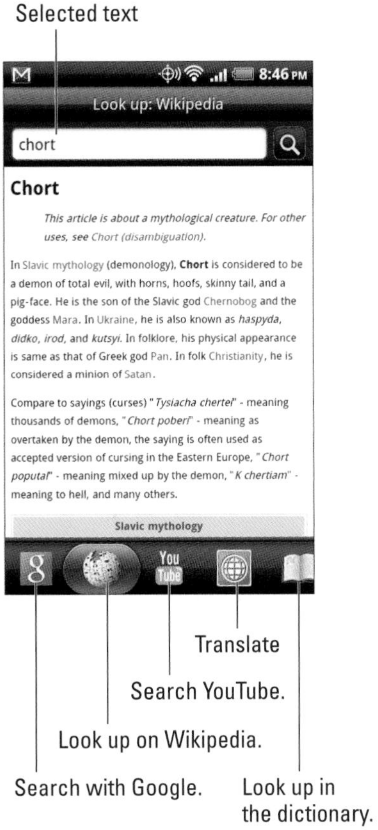

Selected text

Figure 4-8: The Quick Lookup app searches using selected text.

Copying, cutting, and pasting

One of the handiest things you can do with a selected block of text is copy or cut it. Copying and cutting text works on your phone just as it does on a computer. Copying duplicates the text; cutting removes it.

After you select the text, a pop-up slate of icons appears, as shown earlier, in Figure 4-7. To copy the text, choose the Copy icon. To cut the text, choose the Cut icon.

When you cut text, it's removed from the screen.

To paste text, you find a text box or another location where text goes; you can't just paste text anywhere. Odds are good that if you see the onscreen keyboard displayed, you can paste text.

The next step is to summon the Paste command: Long-press the chunk of text until you see the magnifier window. Release your finger and choose the Paste command from the pop-up list of options.

Text can be pasted again and again; there's no limit to how many times you can use the Paste command. The same text is pasted until you copy or cut new text or until you turn off or restart the phone.

Choose Your Keyboard

Three different keyboard layouts are available for you to use on the ThunderBolt: The standard QWERTY keyboard is shown earlier, in Figure 4-1. Figure 4-9 shows the Phone Keypad keyboard, and Figure 4-10 shows the Compact QWERTY keyboard.

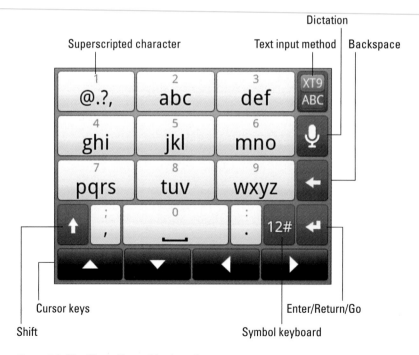

Figure 4-9: The Phone Keypad keyboard.

Shift

Text input method

Superscripted character

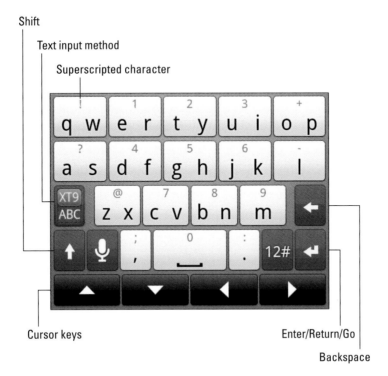

Cursor keys

Enter/Return/Go

Backspace

Figure 4-10: The Compact QWERTY keyboard.

To set which keyboard the ThunderBolt uses, heed these steps:

1. **At the Home screen, press the Menu soft button.**

2. **Choose Settings.**

3. **Choose Language & Keyboard.**

4. **Choose the Touch Input command.**

   You see the Touch Input Settings screen. The top item is used to choose a keyboard, with the current keyboard shown. (I'm guessing that it's QWERTY.)

5. **Choose Keyboard Types.**

6. **Select a keyboard from the list.**

The Phone Keypad and Compact Qwerty keyboards feature alternative input methods, selected by using the XT9 and ABC buttons:

XT9: In this mode, the ThunderBolt interprets your typing, by matching words based on what you type on the keypad. For example, on the Phone Keypad keyboard, you type 43556 and the phone interprets it as *hello.*

ABC: In this mode, you must press each key button a number of times to produce the proper letter. So, on the Phone Keypad, you have to type 44 to get an *H* and then type 33 to get an *E.* Typing *hello* is 44 33 555 555 666. This method may seem awkward, but lots of folks are used to typing on a 9-key phone keypad.

Superscripted characters, shown in Figures 4-9 and 4-10, are produced by pressing and holding keys. In most cases, a pop-up palette of characters appears, from which you can choose which character you want to type.

✔ Other keyboards are available from the Android Market. Some of the good ones, such as Thumb Keyboard, cost money. Others are merely *skins,* which simply change the look of the standard QWERTY keyboard. See Chapter 18 for more information on the Android Market.

✔ Perhaps the most popular alternative keyboard is the Swype keyboard. Swype allows for the rapid input of text because you basically don't lift your finger from the keyboard: You use the Swype keyboard to trace words rather than type them. As this book goes to press, the Swype keyboard variation isn't available for the ThunderBolt. If the situation changes, I'll post an update on my website:

        www.wambooli.com/help/phone

Typing: Use Dictation

The future of communications with your mobile gizmos doesn't involve a keyboard. As a small peek into that future, the ThunderBolt has the capability to interpret your utterances as text. It's a feature I call *dictation.*

The key to using dictation is to locate the Microphone icon, similar to the one shown in the margin. Touch the icon or button and the voice input screen appears, as shown in Figure 4-11.

Figure 4-11: The voice input thing.

Allowing those off-color words

If you really want to use foul, sailorlike or even mother-in-law-like words in your dictation, you need to tell the ThunderBolt that such utterances are allowed. Follow these steps:

1. At the Home screen, press the Menu soft button.

2. Choose Settings.

3. Choose Voice Input & Output.

4. Choose Voice Recognizer Settings.

5. Remove the check mark by the option Block Offensive Words.

There. You're now free to use salty language and have it show up as something other than #### pound signs.

When you see the text *Speak Now,* speak directly into the phone.

As you speak, the Microphone icon (refer to Figure 4-11) flashes. The flashing doesn't mean that the ThunderBolt is embarrassed by what you're saying. No, the flashing merely indicates that the phone is listening, detecting the volume of your voice.

After you stop talking, the phone digests what you said. Eventually, the text you spoke — or a close approximation — appears on the screen. It's magical, and sometimes comical.

✔ The first time you try voice input, you might see a description displayed. Touch the OK button to continue.

✔ A Microphone key is on the onscreen keyboard for when you tire of typing and want to take a stab at some dictation.

✔ The Microphone icon appears only when voice input is allowed. Not every application features voice input as an option.

✔ The better your diction, the better the results. Also, it helps to speak only a sentence or less.

✔ You can edit your voice input just as you edit any text. See the section "Text Editing," earlier in this chapter.

✔ You must speak the punctuation symbols you want in your text. For example, you say, "Sorry I'm late comma William" to have the ThunderBolt produce the text `Sorry about your leg, William` (or similar wording).

✔ Common punctuation symbols you can dictate include the comma, period, exclamation point, question mark, and colon.

✔ Pause your speech before and after speaking punctuation.

✔ The ThunderBolt features a voice censor, which replaces any naughty words you might utter with a series of pound (#) symbols. The ThunderBolt knows a lot of blue terms, including the famous "Seven Words You Can Never Say on Television," but apparently the terms *crap* and *damn* are fine. Don't ask me how much time I spent researching this topic.

Part II
The Phone Thing

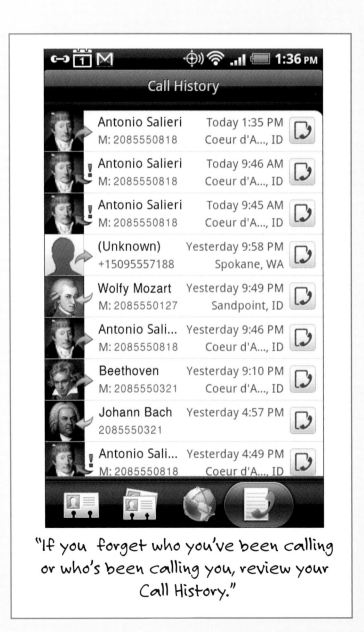

"If you forget who you've been calling
or who's been calling you, review your
Call History."

Colin Carr
Jeremiah Gookin
Jonah Gookin
Jordan Gookin
Simon Gookin

In this part . . .

Imagine how different things would have been if Zeus had used the ThunderBolt for communications. (I'm talking about the phone, not the atmospheric phenomenon.) Sure, having one of the world's first 4G LTE smartphones 3,000 years before anyone else would have been cool — perhaps not as cool as turning yourself into a swan and romancing Tyndareus' wife, but still cool.

The ThunderBolt is certainly a wonderful gizmo, but at its core it does the most basic of cell phone operations: Make phone calls. That's not quite as imaginative or legendary as transmogrification, and it's not the limit of what the phone can do, but it's what's covered in this part of the book.

Antonio Salieri

M 2085550818

(	1	2	3	.	P
)	4	5	6	,	W
+	7	8	9	-	←
␣	*	0	#	$	↵
▲	▼	◄	►		

5

Basic Phone Operations

Did telephones ever come with manuals? I doubt it. For the longest time, the phone was such a simple device. It had a dial or push-button keypad. The phone's receiver was either on or off the hook. You could dial a number and not have to press an Enter key, so the system was smart enough to connect you. It was a thing of wonder.

Your ThunderBolt does far more than just make phone calls, yet it's still capable of that simple operation. Still, the ThunderBolt doesn't have a dial, and there's no keypad on the thing. Therefore, the simple chore of making phone calls requires more documentation than the telephones of yesterday. Rather than give you a boring manual, I present this chapter, which covers the basics of making, receiving, and dealing with phone calls.

Antonio Salieri

M 2085550818
Vienna, Austria

Drag down to answer or
drag up to decline

I Just Called to Say . . .

You can sit around and wait for the phone to ring or you can take the initiative and make that call! All you need is the other person's phone number and a reason to chat. Or, you can be like a teenager and never have a reason to talk but make the attempt anyway.

Well, actually, teenagers prefer to *text* one another, which is covered in Chapter 9, but you get my point.

Making the call

To place a call on your phone, follow these steps:

1. **Touch the phone button, found at the bottom of the Home screen.**

   You see the Phone dialpad, similar to the one shown in Figure 5-1.

2. **Type the number to call.**

   Touch the keys on the dialpad to input the number. If you make a mistake, use the Backspace key, shown in Figure 5-1, to back up and erase.

   As you dial, you may hear the traditional touch-tone sound as you input the number. You might also see matched contacts found on the top part of the phone.

Figure 5-1: Dialing a phone number.

3. **Touch the green Call button to place the call.**

   The phone doesn't make the call until you touch the green button.

   As the phone attempts to make the connection, two things happen:

   - First, the Call in Progress notification icon appears on the status bar and the status bar starts to glow green. The icon and the green glow are your clues that the phone is making a call or is actively connected.

   - Second, the screen changes to show the number you dialed, similar to the one shown in Figure 5-2. When the recipient is in your Contacts list, the name also appears, as shown in the figure. Further, if a picture is part of the person's contact information, the picture appears when the person answers the phone, as shown in Figure 5-2.

Phone number dialed   Contact info

Call in progress   Call duration

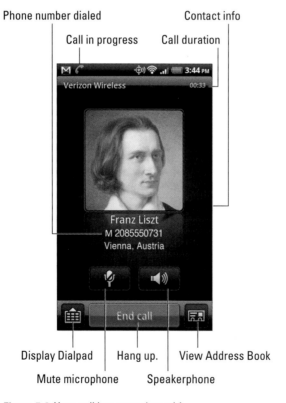

Display Dialpad | Hang up. | View Address Book

Mute microphone   Speakerphone

Figure 5-2: Your call has gone through!

Even though the touchscreen is pretty, at this point you need to listen to the phone: Put it to your ear or listen through the earphones or a Bluetooth headset.

4. When the person answers the phone, talk.

What you say is up to you, though I can recommend from experience that it's a bad idea to open your conversation with your best friend about how you ran over your neighbor's cat — unless you're certain that it's your best friend that you're talking to and not your neighbor.

Use the phone's Volume button (on the side of the ThunderBolt) to adjust the speaker volume during the call.

5. To end the call, touch the red End Call button.

The phone disconnects. You hear a soft *beep,* which is the phone's signal that the call has ended. The Call in Progress notification goes away.

You can do other things while you're making a call on the ThunderBolt. Just press the Home button to run an application, read old email, check an appointment, browse the web, or do whatever. These types of activities don't disconnect you, though your cellular carrier limits the other things you can do with the phone while you're on a call.

You can also listen to music while you're making a call, though I don't recommend it, because the music volume and call volume cannot be set separately.

To return to a call after doing something else, pull down the notifications at the top of the screen and touch the notification for the current call. You return to the Connected screen, similar to the one shown in Figure 5-2. Continue yapping. (See Chapter 3 for information on reviewing notifications.)

- ✔ You can connect or remove the earphones at any time during a call. The call is neither disconnected nor interrupted by doing so.

- ✔ When you're using a Bluetooth headset, I recommend connecting the headset *before* you make the call.

- ✔ If you're using earphones, you can press the phone's Power Lock button during the call to turn off the display and lock the phone. I recommend turning off the display so that you don't accidentally touch the Mute or End button during the call.

- ✔ You can't accidentally mute or end a call when the phone is placed against your face; a sensor in the phone detects when it's close to another object and the touchscreen is automatically disabled.

✔ The phone may vibrate as you dial a number, which is merely a form of feedback. Sound and vibration settings are covered in Chapter 22.

✔ If there's no contact information or picture, a generic android image appears when you make a call. Do not let the robot's image frighten you.

✔ To mute a call, touch the Mute button, shown in Figure 5-2. A Mute icon, shown in the margin, appears as the phone's status (atop the touchscreen).

✔ If you're wading through one of those nasty voice mail systems, touch the Dialpad button (refer to Figure 5-2), so that you can "Press 1 for English" when necessary.

✔ Touch the Speaker button to be able to hold the phone at a distance to listen and talk, which allows you to let others listen and share in the conversation. The Speaker icon appears as the phone's status whenever the speaker is active.

✔ Don't hold the phone to your ear when the speaker is active.

✔ The ThunderBolt has the Flip for Speaker feature. To put the phone in Speaker mode, simply flip the phone over with the touchscreen facing down. The rear speaker takes over.

✔ If you need to dial an international number, press and hold the 0 (zero) key until the plus sign (+) character appears. Then input the rest of the international number. Refer to Chapter 21 for more information on making international calls.

✔ You hear an audio alert when the call is dropped or the other party hangs up on you. The disconnection can be confirmed by looking at the phone, which shows that the call has ended.

✔ You cannot place a phone call when the phone has no service; check the signal strength, as shown earlier, in Figure 5-1. Also see the nearby sidebar, "Signal strength and network nonsense."

✔ You cannot place a phone call when the phone is in Airplane mode. See Chapter 21 for information.

✔ The Call in Progress notification icon (refer to Figure 5-2) is a useful thing. Whenever you see this notification, the phone is connected to another party. To return to the phone screen, swipe down the status bar and touch the phone call's notification. You can then press the End Call button to disconnect or just put the phone to your face to see who's on the line.

✔ Accessing the Internet while making a phone call is possible only when your ThunderBolt is connected to a 4G digital network. You can still access the Internet when using a 3G (or slower) digital connection, but you can't do so while you're making a phone call.

TECHNICAL STUFF

Signal strength and network nonsense

Two peculiar status icons appear to the left of the current time atop the ThunderBolt screen. These icons represent the network the phone is connected to and the signal strength.

The Signal Strength icon displays the familiar bars, rising from left to right. The more bars you see, the better the signal. A very low signal is shown by zero bars; when there's no signal, you see an X over the bars.

When the phone is out of its service area but still receiving a signal, you see the Roaming icon, where an *R* appears near the bars. See Chapter 21 for more information on roaming.

To the left of the Signal Bar icon is the Network icon. No icon means that no network is available, which happens whenever the network is down or you're out of range. The icon might also disappear when you're making a call. Otherwise, you see an icon representing one of the different types of cellular data networks to which the ThunderBolt can connect:

✔ The 1xRTT icon appears when the ThunderBolt is connected to the slow data network. It's better than nothing.

✔ The 3G icon appears when the ThunderBolt is connected to a 3G network. (Figure 5-1 has the 3G icon on display.)

✔ The 4G icon shows its face when the ThunderBolt is communicating with the überfast 4G network.

The Network icon animates whenever a signal is being transmitted.

Also see Chapter 19 for more information on the network connection and how it plays a role in your phone's Internet access.

Dialing a contact

Because your ThunderBolt is also your digital phone book, one of the easiest methods for placing a phone call is to simply dial one of the folks on your People list. You have several ways to do it.

Choose a contact from your People list

To phone up someone in your phone's address book, follow these steps:

1. **On the Home screen, touch the People app's icon.**

   The icon appears at the bottom of the main Home screen, just above the Phone button. After touching the icon, you see a list of contacts. Unless you've messed with the Contacts list, it's sorted alphabetically by first name, similar to the one shown in Figure 5-3.

Contacts

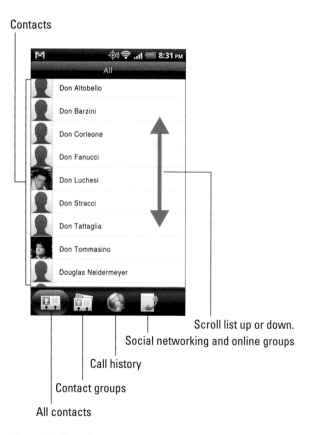

Scroll list up or down.

Social networking and online groups

Call history

Contact groups

All contacts

Figure 5-3: Perusing your people.

2. **Scroll the list of contacts to find the person you want to call.**

   To rapidly scroll, you can swipe the list with your finger or use the tab that appears on the right side of the list, as shown in Figure 5-3; drag the tab around using your finger.

3. **Touch the contact you want to call.**

   The contact's detailed information appears.

4. **Choose the contact's phone number.**

   The contact's information is organized into areas. Phone numbers are found in the Action area: Touch the number to dial.

At this point, dialing proceeds as described earlier in this chapter.

   ✔ You can quickly access the People list when dialing the phone: Touch the People button, illustrated earlier, in Figure 5-1.

   ✔ See Chapter 8 for more information about the People app.

Dial a contact stuck to the Home screen

You have several ways to stick a contact on the Home screen. The most obvious way is to use the Favorites widget, which can be found on the first Home screen to the right of the main Home screen (assuming that you haven't removed the widget since you first set up your ThunderBolt).

The Favorites widget displays contacts you've added to the Favorites group. The widget is configured to show the first nine members of that group in a *Brady Bunch* or *Hollywood Squares* pattern. To access a person, touch their icon in the grid and you see contact information. You can then call the person, as described in Step 4 in the preceding section.

Another way contacts are stuck to the Home screen is by a direct-dial shortcut. Whenever you touch a direct-dial shortcut, the contact is dialed instantly. That's because the shortcut is configured so that the contact's phone number is already chosen.

   ✔ See Chapter 22 for more information about widgets and shortcuts.

   ✔ Chapter 8 describes how to create a direct-dial shortcut on the Home screen as well as how to use the Favorites group.

Phoning someone you call often

Because the ThunderBolt is sort of a computer, it keeps track of your phone calls. Also, you can flag as favorites certain people whose numbers you want to keep handy. You can take advantage of these two features to quickly call the people you phone most often or to redial a number.

To use the call history to return a call or to call someone right back, follow these steps:

1. **Touch the Phone icon on the Home screen.**

   The main dialing screen shows, at the top of the screen, the most recent few calls you've placed (refer to Figure 5-1).

2. **Touch an item in the list to immediately call that person back.**

   If you don't see the person's name and you know that they called recently (or you called them or you missed the call), touch the Hide Dialpad button, also shown in Figure 5-1. You can then pluck the person's name from the list.

People you identify as favorites can be called by scrolling down the Recent Call list to see the favorites group. Heed these directions:

1. **Touch the Phone icon on the Home screen.**

2. **Touch the Hide Dialpad button.**

3. **Scroll down the list a tad.**

   Immediately below the list of recent calls, you see the favorites group. A heavy, black line appears between the groups.

4. **Touch a person's entry to immediately call that person.**

If you keep scrolling the list, you eventually see all contacts in your phone, though that's not this section's point.

Refer to Chapter 8 for information on how to make one of your contacts a favorite.

Where is that call coming from?

The ThunderBolt displays the caller's location for both incoming and outgoing calls, similar to the one shown in Figures 5-2 and 5-4. This feature happens courtesy of the City ID app.

City ID is a subscription service, though you get to use it free for 15 days as a trial right after you activate your ThunderBolt. After that, you have to sign up and pay for the service. Though this tool may not help you identify callers you know, it's handy for gleaning information about unknown incoming calls.

Open the City ID app on the All Apps screen find more information about City ID.

It's for You!

Incoming phone calls are an occasion. Even people who complain about receiving too many phone calls — you know, the types who scowl when they look at the Caller ID to see who's calling — they also enjoy getting phone calls. Even wrong numbers can be exciting.

Receiving a call

Several things can happen when you receive a phone call on your ThunderBolt:

- ✔ The phone rings or makes a noise signaling you to an incoming call.
- ✔ The phone vibrates.
- ✔ The touchscreen reveals information about the call (see Figure 5-4).
- ✔ It turns out to be the President of the United States on the line and you have to stand up, which presents an interesting problem if you're driving a car.

Contact info (if available)

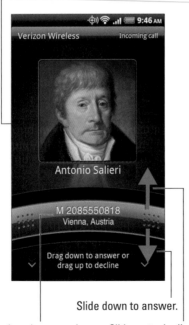

Slide down to answer.

Incoming phone number   Slide up to decline.

Figure 5-4: You have an incoming call.

The last item in the list happens only in the movies. The other three possibilities, or a combination thereof, are your signals that you have an incoming call. A simple look at the touchscreen tells you more information, as illustrated in Figure 5-4.

Figure 5-4 shows what an incoming call looks like when the ThunderBolt is locked. When that happens, slide the lock bar down to answer the call — just as though you were unlocking the phone.

When you're using your ThunderBolt and a call comes in, the phone stops doing whatever you're doing and you see a screen similar to the one in Figure 5-4, but with two buttons and no lock bar: Touch the green Answer button to accept the call.

After answering the call, place the phone to your ear to talk and listen. Or, you can use a headset, if one is attached to the phone (and your head).

To ignore the call, slide the lock bar upward (refer to Figure 5-4) or, if you're using the phone, touch the red Decline button. The phone stops ringing and the call is immediately banished into voice mail.

You can also touch the Volume (Up or Down) button to silence the ringer.

- ✓ The contact picture, such as Mr. Salieri in Figure 5-4, appears only when you've assigned a picture to the contact. Otherwise, the generic Android icon shows up.

- ✓ See Chapter 6 for information on how to deal with an incoming call when you're already on the phone.

- ✓ If you're using a Bluetooth headset, you touch the control on the headset to answer your phone. See Chapter 19 for more information on using Bluetooth gizmos.

- ✓ The sound you hear when the phone rings is known as the *ringtone*. You can configure the ThunderBolt to play a number of ringtones, depending on who is calling, or you can set a universal ringtone. Ringtones are covered in Chapter 6.

- ✓ I don't know whether it's truly a protocol to stand up whenever the president calls, but it seems like it happens a lot in the movies. Do you also notice in movies and television shows that few people say "Goodbye" or any other type of farewell before they hang up the phone? Hollywood. . . .

Setting the incoming call volume

Whether the phone rings, vibrates, or explodes depends on how you configure the ThunderBolt to signal you for an incoming call. Abide by these steps to set the various options (but not explosions) for your phone:

1. **On the Home screen, touch the All Apps button to view all apps on the phone.**

2. **Choose the Settings icon to open the phone's Settings screen.**

3. **Choose Sound.**

4. **Set the phone's ringer volume by touching Volume.**

5. **Manipulate the Ringtone slider to the left or right to specify how loudly the phone rings for an incoming call.**

   After you release the slider, you hear an example of how loudly the phone rings.

6. **Touch OK to set the ringer volume.**

   Meanwhile, back at the main Sound Settings screen:

7. **To activate vibration when the phone rings, touch the gray square by Vibrate to put a green check mark there.**

   With the Vibrate option set, the phone also vibrates on an incoming call.

8. **Press the Home soft button when you're done making settings.**

When the next call comes in, the phone alerts you using the volume setting or vibration options you've just set.

If you'd rather have the phone only vibrate or simply remain silent, choose Sound Profile from the Sound Settings screen. In the Choose a Profile window, select an option to configure how the phone reacts to an incoming call: Normal to have the phone ring or vibrate; Vibrate for vibrate only; or Silent to have the phone just sit there when a call comes in. (Well, you still see the incoming call information on the touchscreen.)

  ✐ Turning on vibration puts an extra drain on the battery. See Chapter 23 for more information on power management for your phone.

  ✐ Also refer to Chapter 22 for additional sound options on the ThunderBolt.

All Your Phone Calls

Unlike your typical teenager or houseguest, the ThunderBolt remembers all the calls you recently received, whether you were there to take them or not. Yes, every single detail is remembered: who called, when they called, whether you answered — all that stuff. The key is knowing where to find the information.

Dealing with a missed call

Missed calls are flagged immediately upon waking the ThunderBolt. Just below the lock bar, you see information about any calls that came in that you didn't catch. If the call was from someone in your People list, you see that person's contact information as well; otherwise, only the number shows up, or a generic `Missed calls` message.

The ThunderBolt doesn't consider a call you've dismissed as being "missed." If you decline the call, as described earlier in this chapter, it isn't an official missed call.

Missed calls are also flagged by an obvious notification icon at the top of the screen, as shown in the margin. To deal with the missed call, pull down the list of notifications and touch the Missed Call notification to see the Call History list.

The Call History list (covered in the next section) shows who called and when. If the missed call was from a contact, you see additional information as saved in your People list.

To call the person back, simply touch the entry in the Call History list, as shown in Figure 5-5.

If you're using a set of earphones with a doodle button, you can press the button twice to have the ThunderBolt redial the most recent call. So if you missed a call, press twice to call the person back.

  ✔ The click-click-doodle-button trick works at any time, not just for a missed call: Pressing the button twice redials the most recent number you dialed, or it dials the number of the last call. See Chapter 1 for more information on what a doodle button is.

  ✔ See Chapter 3 for details on how to deal with notifications.

Reviewing the Call History list

The ThunderBolt keeps a record of all calls you make, incoming calls, and missed calls. Everything is shown on the Call History list (refer to Figure 5-5). To see the list, touch the People app on the Home screen and then touch the Call History button, shown in Figure 5-5.

You can also see the top part of the Call History list when you touch the Phone button. Hide the dialpad to see the rest of the list.

The Call History list shows people who have phoned you or whom you have called, starting with the most recent call at the top of the list. An icon next to each entry describes whether the call was incoming, outgoing, or missed, as illustrated in Figure 5-5.

Who called and whom you called

View contact's call history.

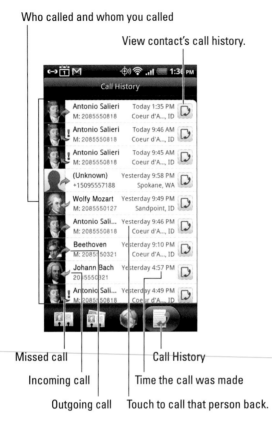

Missed call

Incoming call

Outgoing call

Call History

Time the call was made

Touch to call that person back.

Figure 5-5: The Call History list.

Touching an item in the call log directs the phone to dial the number that was called or missed. To see the person's contact information, long-press the entry and choose the View Contact command.

The call history can become quite long. Use your finger to scroll the list.

 Using the call log is a quick way to add a recent caller as a contact. Simply long-press the entry and choose the command Save to People from the pop-up menu. See Chapter 8 for more information about your phone's address book and the People app.

menu To clear the call log, press the Menu soft button. Choose the Delete All command and touch the OK button to confirm. All entries are then removed from the call log.

Advanced Phone Calls

In This Chapter

▶ Setting up speed dial

▶ Dealing with multiple incoming calls

▶ Making a conference call

▶ Configuring call forwarding options

▶ Banishing a contact to voice mail

▶ Choosing a better ringtone

▶ Assigning ringtones to your contacts

▶ Using your favorite song or sound as a ringtone

*O*nce upon a time, callers heard something called a *busy signal.* I'm not making this up, kids: When you phoned someone who was already talking on the phone, you heard a repetitive drone. *Bonk, bonk, bonk.* It was the busy signal, and it meant that someone was on the line and you had to call back. Yes, there was no call waiting, call forwarding, voice mail, conference calling — none of it.

Your ThunderBolt has the capability to deal with more than one call at a time, set up speed dial, and switch between two or more calls, plus all sorts of fun can be had with ringtones. In a fit of irrational nostalgia, you can even make the old Bell System busy signal sound as your cell phone's ringtone, if you like.

Joseph II
2085550318

Speed Dial Setup

The faster I try to dial a phone, the more random the phone number. Having to retype a phone number isn't the reason I want to dial it quickly in the first place. The solution? Speed dial.

Using speed dial, you can long-press a single digit on the dialpad (refer to Figure 5-1, in Chapter 5). When you release your finger, the speed-dial number is dialed.

To set up a speed-dial number on your ThunderBolt, follow these steps:

1. **From the Home screen, touch the Phone button.**

2. **Press the Menu soft button.**

3. **Choose Speed Dial.**

   The first speed-dial number is already configured to your carrier's voice mail system. The remaining numbers, 2 through 9, are blank.

4. **Touch the Add New button.**

5. **Choose a contact to speed-dial.**

   You can scroll the list or type the first few characters of the contact's name in the Select Contact box.

6. **Choose the contact's phone number to dial from the Number menu button.**

7. **Choose the number on the dialpad to use from the Location menu button.**

   Unused numbers are shown as (Available) in the list. If you choose a number that's already taken, the new contact replaces the original speed-dial contact.

8. **Touch the Save button.**

   The speed-dial number is assigned.

To try out the number, long-press the digit to which you assigned the speed-dial contact.

You see no visual clue on the phone's dialpad that a speed-dial number has been assigned. (The Voicemail icon appears on the 1 key, but you can't change its key or icon.) You can review which speed-dial numbers are assigned to which contacts by repeating Steps 1 through 3, which displays the list.

To remove a speed-dial number, follow Steps 1 through 3 in this section. Long-press the entry you want to remove and then choose the Delete command from the pop-up menu. The speed-dial entry is gone.

✓ Removing a speed-dial entry doesn't delete the contact or the contact's information from the phone.

✓ You can also set a speed-dial number for a contact when you view the contact's information in the People app: Press the Menu soft button while viewing the contact, choose More, and then choose Set Speed Dial. See Chapter 8 for more information on the People app.

✓ You cannot remove speed-dial entry number 1, for voice mail. See Chapter 7 for more information on voice mail for your ThunderBolt.

More than One Call at a Time

They say that humans have trouble juggling more than 9 objects at a time. The record for number of juggled objects is something like 13, though I don't think it's been verified.

Your ThunderBolt is capable of juggling, though I wouldn't toss the phone into the air. When it comes to juggling things, what the ThunderBolt does best is juggle calls. You can receive a call when you're on the phone, add a call, and bounce between calls, all described in this main section.

Getting a call when you're on the phone

You have to be paying attention when you get an incoming call while you're on the phone. You hear a faint beep and perhaps the phone vibrates. To confirm that a call is incoming, and that you aren't imagining things, look at the phone. When someone is calling you while you're already on the line, you see a screen similar to the one shown in Figure 6-1.

You have three options:

Answer the call. Touch the green Answer button to answer the incoming call. The call you're on is placed on hold.

Send the call directly to voice mail. Touch the Ignore button. The incoming call is sent directly to voice mail.

Ignore the call. Do nothing. The call eventually goes to voice mail.

Contact information (if available)

You're already on the phone.

Send the call to voice mail.

Put the current call on hold
and answer the new call.

Incoming phone number

Figure 6-1: Suddenly, there's an incoming call!

When you choose to answer the call and the call you're on is placed on hold, you return to the first call when you end the second call. Or, you can manage the multiple calls as described in the next section.

Bouncing between two calls

After you answer a second call, as described in the preceding section, your ThunderBolt is now working with two calls at once. In this particular situation, you can speak with only one person at a time; juggling two calls isn't the same as a conference call.

The multiple-call screen is shown in Figure 6-2. The yellow and green lights indicate which call you're on and which one is one hold, as illustrated in the figure.

Figure 6-2: Juggling two separate calls at a time.

To switch between callers, touch the contact name or phone number to switch to. The other call is put on hold.

To end the current call, such as the call from Joseph II in Figure 6-2, touch the End Call button, just as you normally would. Both calls might appear to have been disconnected, but that's not the case: In a few moments, the call you didn't disconnect "rings" as though the person called you back. They didn't call you back, though: The ThunderBolt is simply returning you to that ongoing conversation.

✔ The number of different calls your phone can handle depends on your carrier. For most of us, it's only two calls at a time. In this case, a third person who calls you either hears a busy signal or is sent right into voice mail.

✔ Obviously, from the design of the multiple call screen (refer to Figure 6-2), the ThunderBolt may be capable of juggling more than two calls.

✔ Put the phone where you can see the touchscreen when you work with multiple calls. That way, you can see who's on the line, who's waiting, and how long both callers have been waiting.

✔ If the person on hold hangs up, you may hear a sound or feel the phone vibrate when the call is dropped.

Configuring a conference call

A *conference call* is one that you set up, where you call one person and then add another person to the conversation. Unlike juggling multiple calls, a conference call is something you do intentionally.

You create a conference call by phoning the first person and then adding another person to the call. Here's how it works:

1. **Phone the first person.**

2. **After your phone connects and you exchange pleasantries, press the Menu soft button.**

3. **Choose the Add Call command.**

   The first person is put on hold.

4. **Dial the second person.**

   You can use the dialpad or choose the second person from your People list or Recent call log.

   Say your pleasantries and inform the party that the call is about to be merged.

5. **Touch the Merge Calls button.**

   The two calls are now joined: You see both contacts displayed on the screen, though no one is on hold. Everyone is talking!

6. **Touch the End Call button to end the conference call.**

   All calls are disconnected.

When several people are in a room and want to participate in a call, you can always put the phone in Speaker mode: Touch the Speaker button.

Send That Call Somewhere Else

No law says that you have to answer each and every incoming call. My personal rule is not to answer any unknown numbers, unless someone owes me money. Beyond that, it's simple on the ThunderBolt to banish any incoming call, either to dismiss it straight to voice mail or to send the call off to another phone number using the call forwarding technique.

Forwarding phone calls

The process by which you take a phone call coming into your ThunderBolt and send it elsewhere is known as *call forwarding*. For example, you can send all calls you receive to your office when you're on vacation. Then you have the luxury of having your cell phone and still making calls but being blissfully ignorant of anyone calling you.

The options for call forwarding on the ThunderBolt are set by the cell phone carrier, and not by the phone itself. In the United States, using Verizon as your cellular provider, the call forwarding options work as described in Table 6-1.

Table 6-1	Verizon Call Forwarding Commands	
To Do This	*Dial These Characters*	*Followed By This*
Forward unanswered incoming calls	*71	The forwarding number
Forward all incoming calls	*72	The forwarding number
Cancel call forwarding	*73	[Nothing]

For example, to forward all calls to (714) 555-4565, you input ***727145554565** and touch the green Dial button on the ThunderBolt. You hear just a brief tone after dialing, and then the call ends. After that, any call coming into your phone rings at the other number.

▸ You must disable call forwarding on your ThunderBolt to return to normal cell phone operations. Dial *73.

▸ The ThunderBolt doesn't even ring when you forward a call using *72. Only the phone number you've chosen to forward to rings.

▸ You don't need to input the area code for the forwarding number when it's a local call. In other words, if you only need to dial 555-4565 to call the forwarding number, you need to input only ***725554565** to forward your calls.

✔ If you're using Google Voice as your voice mail service, forwarding all calls on your ThunderBolt also affects Google Voice. Specifically, you need to reconfigure Google Voice after you cancel call forwarding. See Chapter 7 for more information on Google Voice and voice mail.

Redirecting someone to voice mail

Here's a neat trick: Configure the ThunderBolt to forward an incoming call to voice mail if that call is from one of your contacts and you've directed the phone to persistently banish the person to voice mail. Follow these steps:

1. **Touch the People icon on the Home screen.**

   The People app opens.

2. **Choose a contact.**

   Use your finger to scroll the list of contacts until you find the annoying person you want to eternally banish to voice mail.

3. **If necessary, scroll to the bottom of the screen to see the Options area for the contact.**

4. **Choose Block Caller.**

   A confirmation warning appears.

5. **Touch the OK button.**

   The contact is now configured so that all incoming calls from that person — from their phone numbers listed in the contact's information — are immediately sent to voice mail. Your ThunderBolt won't even ring.

To unbanish the contact, repeat these steps. Confirm that the contact isn't blocked, by looking for the text *Off* beneath the Block Caller item.

✔ This feature is one reason you might want to retain contact information for someone with whom you never want to have contact.

✔ See Chapter 8 for more information on contacts.

✔ Also see Chapter 7.

Ringtone Mania

Your phone doesn't just ring. No, it has a *ringtone,* a special sound you can choose, create, or customize. It can be the latest tune from the Red Hot Chili Peppers or one from your childhood that would annoy everyone when it plays.

✔ Ringtones aren't just for phone calls: You can also assign ringtones to notifications (such as when you get a new email message), calendar appointments, alarms, and other noises made by the ThunderBolt.

✔ You can assign ringtones to individual contacts, which helps you know who's calling just by hearing the phone ring.

Changing the ringtone

To select a new ringtone for your phone, or to simply confirm which ringtone you're using already, follow these steps:

1. **From the Home screen, touch the All Apps button.**

2. **Choose Settings.**

3. **Choose Sound.**

4. **Choose Phone Ringtone.**

   If you have a ringtone application, you may see a menu that asks you which source to use for the phone's ringtone. Choose Android System.

5. **Choose a ringtone from the list that's displayed.**

   Scroll the list. Tap a ringtone to hear a preview.

6. **Touch Apply to set the ringtone, or press the Back soft button to cancel and keep the phone's ringtone as is.**

To set the ringtone used for notifications: In Step 4, choose Notification Sound rather than Phone Ringtone.

Setting a contact's ringtone

To assign a specific ringtone to a contact, follow these steps:

1. **Touch the People app icon on the Home screen.**

2. **From the list, choose the contact to which you want to assign a ringtone.**

3. **Scroll down the contact's information until you see the Options area.**

4. **Choose Ringtone.**

5. **Choose a ringtone from the list.**

   It's the same list that's displayed for the phone's ringtones.

6. **Touch OK to assign the ringtone to the contact.**

Whenever that contact calls you, the ThunderBolt rings using the ringtone you specified. That way, you can determine that Larry is calling and know that it's Larry without even looking at the phone.

To remove a specific ringtone for a contact, repeat the steps in this section but choose the ringtone named Default Ringtone. This option sets the contact's ringtone to be the same as the phone's ringtone.

Using music as a ringtone

You can pull any tune from the ThunderBolt music library and use that song as the phone's ringtone. The first step, of course, is to get music into your phone: See Chapter 16 for details on how to use your ThunderBolt as a portable MP3 music player.

After you stock the phone with tunes, follow along with these steps:

1. **At the Home screen, press the Menu soft button.**
2. **Choose Settings.**
3. **Choose Sound.**
4. **Choose Phone Ringtone.**
5. **Touch the New Ringtone button atop the list of ringtones.**

   Instantly, you see the collection of tunes you have in the phone. All songs are arranged alphabetically.

6. **To choose a song to use as a ringtone, touch the circle by the song's title.**

   The song plays; touch the song again to stop.

7. **Touch the OK button to add the tune to your ringtone collection.**
8. **Touch the Apply button to assign the song as the ThunderBolt's main ringtone sound.**

As a ringtone, the song plays — from the start of the song — when you have an incoming call and until you answer the phone, send the call to voice mail, or choose to ignore the call and eventually the caller goes away and the music stops.

  ✔ You can add as many songs as you like to the ThunderBolt's ringtone repertoire by repeating the preceding steps.

  ✔ Follow the steps in the earlier section "Changing the ringtone" for information on switching between different song ringtones.

- ✔ Refer to the steps in the earlier section "Setting a contact's ringtone" to assign a specific song to a contact.

- ✔ A free app at the Android Market, Zedge, has oodles of free ringtones available for preview and download, all shared by Android users around the world. See Chapter 18 for information about the Android Market and how to download and install apps such as Zedge on your phone.

Creating your own ringtones

You can easily make your own ringtones. All you need is either an MP3 or a WAV audio file, such as a personalized message, a sound you record on your computer, or a sound tidbit you stole from the Internet. As long as the sound is in the MP3 or WAV format, it can be used as a ringtone on your phone.

The secret to creating your own ringtone is to transfer the audio file from your computer to the ThunderBolt. This topic is covered in Chapter 20, on synchronizing music between your computer and phone.

After the audio file is on your phone, it's placed in the music library. After it's there, you can choose the audio clip as a ringtone in the same way you can assign any music on the ThunderBolt as a ringtone, as described in the preceding section. The file will most likely show up under the title Unknown Album or Unknown Artist.

Please Leave a Message

*I*t took me a while to figure out the difference between having an answering machine and having voice mail. There is no difference, of course. *Voice mail* is simply a fancier term for having an answering machine. Or, voice mail can imply some kind of technology that's far more advanced than a tape recorder hooked up to your telephone. It's as though voice mail technology came from the future, perhaps a technology stolen from the UFO that landed in Roswell, New Mexico, in 1947. Yep, if voice mail is good enough for extraterrestrials, it's good enough for you.

cemail

Joseph II
(208) 555-0131
Received: 9:10pm (7 secs)

I can't find your composition, please res◄

Thank you.

Boring, Boring Carrier Voice Mail

The basic form of voice mail, and therefore the least sophisticated and most stupid, is the voice mail service provided by your cellular provider. It has no fancy features, and nothing special. So don't read the later parts of this chapter about the newer, better types of voice mail lest you be sorely disappointed.

Carrier voice mail picks up missed calls and calls you thrust into voice mail. The ThunderBolt alerts you to a missed call by displaying the Missed Call notification (shown in the margin). You then dial the voice mail system, listen to your calls, and use the phone's dialpad to delete messages or repeat messages or use other features you probably don't know about because no one ever pays attention.

- ✔ The Missed Call icon may not appear when you've chosen to ignore the call or pressed the Decline button to send a call to voice mail.
- ✔ You cannot use Visual Voice Mail, covered later in this chapter, until you've set up basic voice mail on your phone, as described in this section.

Setting up carrier voice mail

If you haven't yet done it, you need to set up voice mail on your phone. Even if you believe it to be set up and configured, consider churning through these steps, just to be sure:

menu

1. **From the Home screen, press the Menu soft button.**

2. **Choose Settings.**

   The Settings screen appears.

3. **Choose Call.**

   Voice mail options are set in the Voicemail section of the Call screen.

4. **Choose Voicemail Service.**

5. **Choose My Carrier, if it isn't chosen already.**

After you choose a voice mail service, you see the Voicemail Settings and Clear Voicemail Notification items appear on the Call screen.

You can use the Voicemail Settings command to confirm or change the voice mail phone number. For Verizon in the United States, the number is *86.

After performing the preceding steps, the next step is to call into the carrier's voice mail service to complete the setup: Dial *86. On my ThunderBolt, I configured my language, set a voice mail password, and then recorded a greeting, following the steps offered by the Cheerful Verizon Robot. Complete these steps even if you plan to use an alternative voice mail service, covered elsewhere in this chapter.

Don't forget to complete your voice mailbox setup by creating a customized greeting. If you don't, you may not receive voice mail messages, or people may believe that they've dialed the wrong number.

Retrieving your messages

When you have pending voice mail, you see a message on the ThunderBolt's unlock screen. it tells you how many voice mail messages you have, plus the times the messages were left.

After unlocking the phone, you can choose the New Voicemail notification, shown in the margin, which automatically dials into the carrier voice mail system. Type your voice mail password and listen to your messages.

Table 7-1 lists the commands for using Verizon voice mail service (current at the time this book went to press). These commands may change later, though the Cheerful Verizon Robot reminds you of the commands while you're using carrier voice mail.

Table 7-1	Verizon Voice-Mail System Commands
Dial This Character	*To Do This*
*	Go to the main menu or, if you're at the main menu already, disconnect from voice mail
0	Get help
1	Listen to messages or rewind
2	Send a message to another phone number on the Verizon system
3	Fast-forward
4	Review or change your personal options, such as the message greeting
5	Restart the session or get date-and-time information on a message
6	Send the message to someone else
7	Delete the message you just heard
8	Reply to the message
9	Save the message you just heard
19	Review erased messages
88	After listening to a message, call the sender
#	End input

TIP

- The Voicemail app on the All Apps screen can also be used to dial into the Verizon Voicemail service. The app is also used to set up and access Visual Voice Mail, covered in the next section.

- The voice mail number is preset on your ThunderBolt to speed-dial the number 1: Whenever you see the phone dialpad, press and hold the 1 key to quickly access carrier voice mail.

- You don't have to venture into carrier voice mail just to see who's called you. Instead, check the call log to review recent calls. Refer to Chapter 5 for information on reviewing the call log.

- The New Voicemail notification icon doesn't go away just because you called into the voice mail service. Nope, you have to delete any new messages to remove the notification icon.

- Calls you exile into voice mail aren't flagged as Missed in the Recent call log.

- See Chapter 3 for more information on reviewing notifications.

Visual Voice Mail

A less boring option than carrier voice mail is something called Visual Voice Mail. It's an app that lets you organize and listen to your messages in an interactive way. The only drawback to using Visual Voice Mail is that it costs extra. You must subscribe to the service, which runs $2.99 per month at the time this book goes to press.

Setting up Visual Voice Mail

To configure Visual Voice Mail to work on your ThunderBolt, first set up carrier voice mail as covered earlier in this chapter. Visual Voice Mail is simply an interface into your existing carrier voice mail.

After you get carrier voice mail up and running, and especially after you set your password or PIN, follow these steps:

1. **Touch the All Apps button on the Home screen.**

   A list of all apps installed on your phone appears.

2. **Choose the Voicemail app.**

3. **Touch the Visual Voice Mail item atop the screen.**

4. **Type your current carrier voice mail password.**

5. **Touch the Sign In button.**

6. **Touch the Subscribe link to complete the setup.**

Or, you can touch the Exit button to chicken out.

The Voicemail app may need upgrading before you can access Visual Voice Mail. If so, touch the Upgrade button and use the Android Market app to upgrade and install the Visual Voice Mail program. After the program upgrade is installed, run the Voicemail app and follow the directions on the touchscreen to set things up.

Refer to Chapter 18 for more information on using the Android Market to install new applications on your ThunderBolt.

Accessing Visual Voice Mail

Visual Voice Mail serves as your access to all your voice mail. After Visual Voice Mail is configured (see the preceding section), you never again need to dial carrier voice mail. Simply pull down a Visual Voice Mail notification or start the Voicemail app, and all your messages are instantly available on the screen.

Touch an item in the Voicemail inbox to review the message. Use the controls on the screen to review your message, call the person back, or delete the message.

Visual Voice Mail uses the same greeting that was set when you first configured carrier voice mail. To change the greeting, you have to dial carrier voice mail and follow the menus.

The Wonders of Google Voice

Perhaps the best option I've found for working your voice mail is something called Google Voice. It's more than just a voice mail system: You can use Google Voice to make phone calls in the United States, place cheap international calls, and perform other amazing feats. But for the purposes of this section, the topic is using Google Voice as the ThunderBolt's voice mail system.

✓ Even when you choose to use Google Voice as your ThunderBolt's voice mail service, I still recommend setting up and configuring basic, boring carrier voice mail as covered earlier in this chapter.

✔ Call forwarding can affect Google Voice. You may need to reset Google Voice after using call forwarding. See Chapter 6 for more information on call forwarding, and see the next section "Setting up a Google Voice account" for information on reestablishing Google Voice as your phone's voice mail service. Oh, here it is:

Setting up a Google Voice account

I recommend getting a Google Voice account on the Internet before you configure the ThunderBolt for Google Voice. Start your adventure by visiting the Google Voice home page on the Internet:

```
http://voice.google.com
```

If necessary, sign in to your Google account. It's the same account name and password you use to access your Gmail.

Your next task is to configure a Google Voice number to be used for your ThunderBolt, as covered in the next section.

✔ If all you want is to use Google Voice as your voice mail service, choose the option Just Want Voicemail for Your Cell.

✔ Google Voice offers a host of features, including international dialing, call forwarding, and other stuff I'm not aware of.

Adding your ThunderBolt to Google Voice

After you have a Google Voice account, you need to add the ThunderBolt's phone number to the list of phone numbers registered for Google Voice. As in the preceding section, I recommend that you complete these steps on a computer connected to the Internet, but keep your phone handy:

1. **Click the Gear icon in the upper right corner of the Google Voice home page.**

   You need to access the Voice Settings command, which may change its location in a future update to Google Voice.

2. **Choose the Voice Settings command.**

   You need to access the Settings screen, where you register phone numbers for use with Google Voice.

3. **Click the link Add Another Phone.**

4. **Work the steps to verify your phone for use with Google Voice.**

Eventually, Google Voice needs to phone up your ThunderBolt. When it does, use the dialpad to type the code number you see on your computer screen.

After confirming the ThunderBolt, you see it listed as a registered phone, though you're not done yet:

5. **Click the Activate Voicemail link.**

6. **On your ThunderBolt, dial the number you see on your computer screen.**

    The number starts with *71, which is the command to forward unanswered calls on your phone. Note that the number you're dialing isn't the same as your Google Voice phone number.

    The number dials and then the ThunderBolt hangs up right away. That's normal.

7. **On your computer screen, click the Done button.**

    The phone is now registered for use with Google Voice.

If you've messed with configuring voice mail on the ThunderBolt, you may have noticed some options available for configuring the phone to work with Google Voice, found on the Call screen (described earlier in this chapter). Don't bother following these steps, because it can be a waste of time. By following the preceding steps, as well as my recommendations in the next section, you'll have a better setup for dealing with Google Voice as your phone's voice mail service.

Getting your Google Voice messages

Google Voice transcribes your voice mail messages. They all show up eventually in your Gmail account, just as though someone sent you an email rather than left you a voice mail. It's a good way to deal with your messages, but it's not the best way.

 The best way to handle Google Voice is to use the Voice app, available from the Android Market. Use the QR code in the margin, or visit the Android Market to search for and install the Google Voice app. (See Chapter 18 for details on how to use the Market.)

After the Google Voice app is installed, it provides the best interface for receiving messages. The interface looks similar to an email program. You can review your messages or touch a message to read or play the message, as illustrated in Figure 7-1.

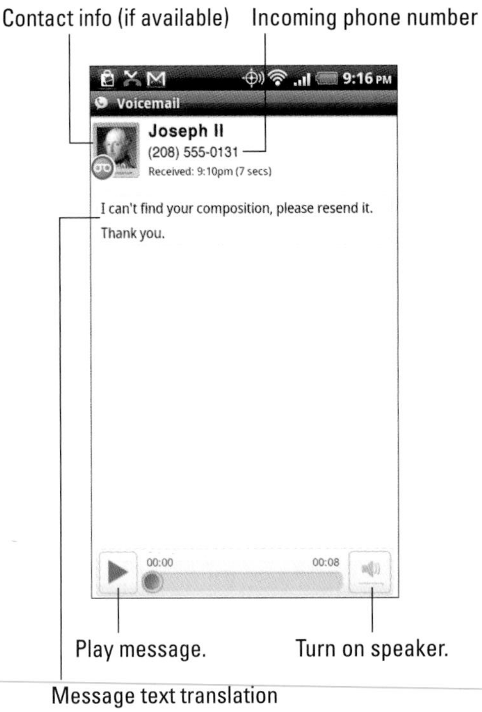

Contact info (if available)   Incoming phone number

> **Voicemail**
>
> **Joseph II**
> (208) 555-0131
> Received: 9:10pm (7 secs)
>
> I can't find your composition, please resend it.
> Thank you.
>
> ▶ 00:00   00:08 🔊

Play message.   Turn on speaker.

Message text translation

Figure 7-1: Voice mail with the Google Voice app.

 When new Google Voice messages come in, you see the Google Voice notification icon appear, as shown in the margin. Pull down the notifications and choose the Voicemail from *Whomever* item to read or listen to the message.

- ✔ With Google Voice installed, you see two notices for each voice mail message that's left: one from Google Voice and another for the Gmail message that comes in.

- ✔ You can best listen to the message when using the Google Voice app. In Gmail, you see a transcript of the message, but you must touch the Play Message link to visit the Internet and then listen to the message.

- ✔ The text translation feature in Google Voice is astonishingly accurate at times, and at other times not so good.

- ✔ The text *Transcript Not Available* appears whenever Google Voice is unable to create a text message from your voice mail or whenever the Google Voice service is temporarily unavailable.

Organize Your Friends

In This Chapter

▶ Exploring the People app

▶ Searching and sorting your contacts

▶ Adding new people

▶ Editing and changing contacts

▶ Linking like contacts

▶ Making a contact group

▶ Deleting contacts

*P*op quiz time! You get 100 *For Dummies* points if you can answer this chapter's question.

Q: *What is a Rolodex?*

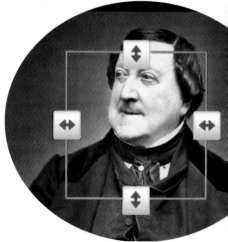

   A. A type of dinosaur.

   B. A fancy watch that you wear around your finger.

   C. An evil minion of the dark lord *Overdex.*

   D. A nostalgic piece of office equipment designed to store information about contacts in a rotating file index. It's no longer necessary to own because most people keep all their contact information — names, phone numbers, addresses, email, and other info — inside handy electronic devices, such as their ThunderBolt phones.

The answer, of course, is D.

Your Phone's Pantheon of People

Your phone is the primary place where you keep information about all the people you know. In fact, it most likely already contains this type of information. That's because your Google account was synchronized with the phone's address book when you first configured the ThunderBolt.

The ThunderBolt address book is accessed from the People app, which is this section's topic.

- ✔ If you haven't yet set up a Google account, refer to Chapter 2.

- ✔ Adding more people to your phone's address book is covered later in this chapter, in the section "Making a new contact."

- ✔ Most apps on the ThunderBolt use information from the People app. These apps include Email, Gmail, Latitude, Messages, and any other app that lets you share information such as photographs or videos.

- ✔ Your social networking friends are also included in the People app's address book. See Chapter 12 for more information on using the ThunderBolt as your social networking hub.

Accessing the address book

To peruse the ThunderBolt address book, run the People app. This app can be found on the main Home screen as well as on the All Apps screen. The address book also lurks behind the dialpad. It's practically ubiquitous!

The People app shows a list of all contacts in your ThunderBolt, as shown in Figure 8-1. Your own account comes first, followed by Verizon shortcut accounts and then the rest of your address book, sorted by first or last name.

Scroll the list by swiping with your finger. Or, as the list scrolls, you can grab the thumb tab that appears on the right side of the screen to quickly scroll.

To do anything with a contact, you first have to choose it: Touch a contact's name and you see detailed information, illustrated in Figure 8-2.

Wierdo Verizon contacts

Your info   Create a new contact.

Matched contact exists

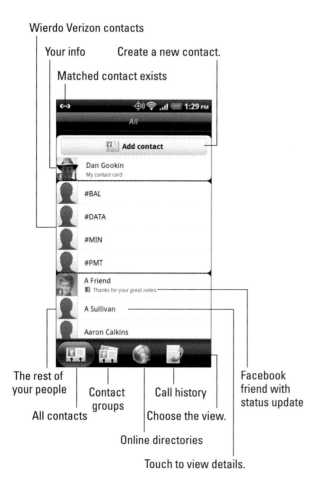

The rest of
your people   Contact   Call history

       groups

    All contacts   Choose the view.

Facebook
friend with
status update

Online directories

Touch to view details.

Figure 8-1: Your ThunderBolt address book.

The information is organized into these areas:

Action: The Action area contains things you can do with the contact, such as call their phone (or phones), send them messages, compose email, view their addresses on a map, visit their social networking sites, or perform other activities.

Information: The Information area lists items such as the contact's birthday, organizations they belong to, and website.

Do things.   View or link accounts.

Contact picture

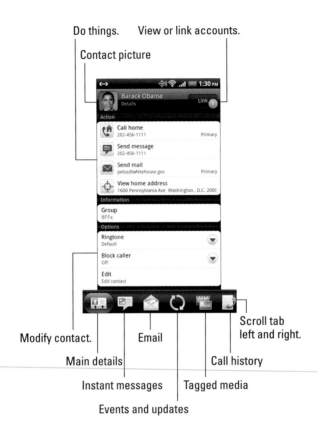

Scroll tab
left and right.

Modify contact.   Email

Main details   Call history

Instant messages   Tagged media

Events and updates

Figure 8-2: Contact information.

Options: The Options area lists commands to change the contact's ringtone, block the caller, or edit the contact information.

 After viewing the contact's detailed information, you can press the Back soft button to return to the main People page.

✔ See the later sidebar "Those weird people" for information on the first group of unusually named contacts in the People app.

✔ See the later section "Matching identical contacts" for information on using the Link button. Basically, this button helps you join two separate contacts that might exist for a single person.

 ✔ Not every phone number can receive text messages. President Obama's home number, shown earlier, in Figure 8-1, is the White House main line, which doesn't accept text messages — well, at least none of the messages I've sent.

✔ After you choose a contact's email address, you see a pop-up menu asking which email program to use: Gmail or another email account.

✔ Not every contact has a picture, and the picture can come from a number of sources (Gmail or Facebook, for example). See the later section "Add a picture" for more information.

✔ Rather than endlessly scroll the address book and run the risk of rubbing your fingers to nubs, you can employ the ThunderBolt's powerful Search command. You do so by touching the Search soft button while using the People app. The Search People text box appears, in which you can type all or part of a contact's name. The more you type, the narrower the list of matching contacts.

✔ There's no correlation between the number of contacts you have and how popular you are in real life.

Sorting the address book

Your contacts are displayed in the People app in a certain order: alphabetically by first name. You can change the order, as well as whether contacts are listed by first name first or last name first. Here's how:

1. **Start the People app.**

menu 2. **Press the Menu soft button.**

3. **Choose the View command.**

   The View Contacts screen appears.

4. **Choose Sort Contact List.**

5. **Choose First Name or Last Name to sort contacts by their first names or last names.**

   The ThunderBolt ships with the First Name option already set.

6. **Choose View Contact Name As.**

7. **Choose First Name First or Last Name First, which specifies how the contacts are displayed in the list.**

   The ThunderBolt ships with the option First Name First selected.

   As long as you're at the View Contacts screen:

8. **Place check marks by all items on the Options menu.**

   Each item represents a location from which the People app can pull contacts for the phone's address book.

9. **Touch the Done button.**

Those weird people

Some people in the People app aren't people at all. They're *preset* contacts, which represent various phone company services. Here's the list:

#BAL: Receive a free text message indicating your current cell phone charges as well as any previous payments you've made.

#DATA: Receive a free text message indicating your text message or data usage.

#MIN: Receive a free text message indicating the minutes you've used on the ThunderBolt, including peak, off-hour, or weekend minutes or whatever other categories for cell phone minutes they can devise.

#PMT: Make a payment using your phone. This operation works only when you've configured your account to make payments via the phone.

Customer Care: Contact Verizon support for your phone. (It's a shortcut for the number 611, which is the support number for your cell phone.)

Warranty Center: Contact Verizon for troubleshooting and warranty issues regarding your ThunderBolt.

See Chapter 9 for more information on text messaging.

There's no right or wrong way to display your contacts — only whichever method you're used to. I prefer them sorted by last name and listed first name first.

Work with Your People

There's a lot you can do with your ThunderBolt address book. Beyond making phone calls, there's email, texting, chat, using the Maps app to locate people and businesses, and a host of other fun and useful things. Various chapters throughout this book cover these tasks. The tasks at hand in this section are to manage those friends, create new contacts, update information, and do a host of other useful things.

Making a new contact

You have many, many ways to put new contacts into your ThunderBolt phone.

Add a contact from a recent call

Perhaps the quickest way to add a new contact is to use the list of recent callers. After someone calls you, or you call them, you can use the Call History feature to create a new contact. Obey these steps:

1. **From the Home screen, touch the People icon.**

2. **Choose the Call History button.**

   The button is at the bottom of the screen, as shown earlier, in Figure 8-1.

3. **Long-press the phone number for which you want to create a new contact.**

   Most likely, it's the top number in the list.

4. **Choose Save to People.**

5. **Choose Create New Contact to make a new contact for the number.**

   You can also choose the Save to Existing Contact command to add the phone number to an existing contact — for example, when Ron finally discloses that second cell phone number he swears she doesn't have, but does anyway. In that case, locate the contact in the list and skip to Step 7.

6. **Fill in the contact's information.**

   Use the onscreen keyboard to type as much information as you know about the contact. Of key importance are the first and last names.

   Use the Up or Down arrow buttons on the onscreen keyboard to hop between the various text fields.

7. **Touch the Save button.**

   The contact has been created.

Make a new contact from scratch

Sometimes it's necessary to create a contact when you meet another human being in the real world. In that case, you have more information to input, and it starts like this:

1. **Open the People app.**

2. **Touch the Add Contact button.**

   Refer to Figure 8-1 for its location.

3. **From the Contact Type button menu, choose Google.**

   When you choose Phone or SIM, the contact is saved only on your ThunderBolt. By choosing Google, you're assured that the contact is backed up to the Internet and available on other Android devices as well as on any computer you use to access your Google account.

4. **If prompted, touch the OK button to confirm that you want to create a Google contact.**

5. **Fill in the information onscreen as best you can.**

Fill in the text fields with the information you know, as shown in Figure 8-3. Use the Delete and Add buttons, as illustrated in the figure, to create more items. For example, if the contact has more than one phone, you'll want to add them all. (You don't need to remove empty fields.)

6. **Touch the Done button to complete editing and add the new contact.**

Figure 8-3: Creating a new contact.

The new contact is automatically synced with your Google account on the Internet — as long as you chose Google in Step 3. That's one beauty of the Android operating system used by the ThunderBolt: You have no need to duplicate your efforts; contacts you create on the phone are instantly synchronized with your Google account on the Internet.

Build a contact on the Internet

Because your ThunderBolt is linked to your Gmail account on the Internet, anytime you add a contact to your list of Gmail contacts, the same contact is also added to your phone. Also, by using Gmail on a computer, it's often easier to input lots of information; having a larger screen and a real keyboard helps.

To add a contact to your Gmail account on your computer, follow these steps:

1. **Log in to your Google Gmail account at** http://gmail.google.com.

2. **Choose Contacts from the links listed on the left side of the page.**

3. **Click the New Contact button.**

4. **Fill in the contact's information.**

   Use the various Add links to type in additional email addresses, phone numbers, and other information. For example, you can specify both home and work email or phone numbers.

5. **If necessary, click the Save button when you're done.**

   Gmail saves automatically, so the Save button may not be available. That's okay; the contact's info is saved.

The new contacts appear on your ThunderBolt almost immediately.

Import contacts from your computer

Your computer's email program is doubtless a useful repository of contacts that you've built up over the years. You can export these contacts from your email program and then import them into the ThunderBolt's People app. It's not the simplest thing, but it's possible.

The key is to save or export the records in the *vCard* (.vcf) file format. In most sophisticated email programs, exporting in the vCard file format is a common feature. Here's how to find it on most popular email programs:

- ✔ In the Windows Live Mail program, choose Go➪Contacts and then choose File➪Export➪Business Card (.VCF) to export the contacts.

- ✔ In Windows Mail, choose File➪Export➪Windows Contacts and then choose vCards (Folder of .VCF Files) from the Export Windows Contacts dialog box. Click the Export button.

- ✔ On the Mac, open the Address Book program and choose File➪Export➪ Export vCard.

After the vCard files are created, connect the ThunderBolt to your computer and transfer the vCard files from your computer to the phone. Directions for making this type of transfer can be found in Chapter 20.

You don't need to copy the vCard files to any specific folder on the ThunderBolt's MicroSD card. Placing them in the root folder is fine.

After the vCard files are on the ThunderBolt, you start the People app and follow these steps to complete the process:

menu

1. **Press the Menu soft button.**

2. **Choose the Import/Export command.**

3. **Choose Import from SD Card.**

4. **Choose the Google option from the Save Contact To menu.**

5. **If you're prompted, choose the option Import All vCard Files.**

   You want to import all the vCard files you copied over.

6. **Touch the OK button.**

   The contacts are saved to your Gmail account, which instantly creates a backup copy.

The importing process may create some duplicates. That's okay: You can link two entries for the same person in the People app. See the section "Matching identical contacts," later in this chapter.

Bring in contacts from your social networking sites

The ThunderBolt is keenly aware of your social networking relationships and how vital they are to keeping your online persona popular. For example, all your Facebook and Twitter friends can quickly be linked into the phone's address book — and probably have been linked in already.

To see your social networking contacts, you use the Online Directories button at the bottom of the screen in the People app. (Refer to Figure 8-1.) Touch the button and then choose a directory.

For example, touch Facebook. If you haven't yet configured social networking on your ThunderBolt (covered in Chapter 12), you're asked to log in to Facebook, or into whichever directory you selected. Otherwise, you see a list of your Facebook friends.

Choosing a contact on the Online Directories screen displays information about that contact as saved on the social networking or other Internet site. You may see other information as well, including status updates or media.

All contacts from your online directories are merged onto the main screen in the People app. And, if you want to see updates or other information, you can always use the Friend Stream app, as covered in Chapter 12, or an individual app for the social networking site.

Find a new contact on the map

When you use the Maps app to locate a restaurant, chiropodist, or bordello — or all three in one place — you can quickly create a contact for that location. Here's how:

1. **After searching for your location, touch the cartoon bubble that appears on the map.**

   You see more details for the location, as shown in Figure 8-4, where a fish restaurant has been found.

Location you searched for

Found location (cartoon bubble)

Figure 8-4: Adding a contact from the Maps app.

2. **Touch the More button.**

   The More button is shown in the margin. It appears on the screen detailing information for the location you found on the map.

3. **Choose Add As a Contact.**

   A new contact screen appears, similar to Figure 8-3. You might find a lot of the information already filled in, which is simply information copied from the Maps app.

4. **Ensure that Google is chosen for the contact type; touch the OK button to confirm.**

   I recommend saving the contact to your Google account because it's automatically synchronized and backed up on the Internet.

5. **Optionally, add more information about the contact, if you know it.**

6. **Touch the Save button.**

   The new contact is created.

See Chapter 13 for detailed information on how to search for a location using the Maps application.

Editing a contact

Sure, some people live at the same place forever and have the same phone number they used back in the 1960s. That person is rare. For the rest of us, contact information changes all the time. You deal with this process on the ThunderBolt by editing and updating your contact information.

Basic changes

Minor corrections and additions can be made to any contact by locating and displaying the contact's information:

1. **Choose the contact's name from the People app.**

2. **Press the Menu soft button.**

3. **Choose Edit.**

4. **Make any changes as you see fit.**

   Details on making specific changes appear in the sections that follow.

5. **Touch the Save button when you're finished editing.**

Use the Delete and Add buttons by a field (refer to Figure 8-3) to add and remove items for the contact.

When editing a matched contact, you see multiple sets of information. For example, a contact linked between Google and your ThunderBolt will have separate entries — one for Google and another for Phone. Both sets of information are displayed for the contact, but when you make changes, the changes are applied to only one source at a time.

See the later section "Matching identical contacts" for more information on linking contacts.

Add a phone number

When you enter or edit a phone number, the ThunderBolt displays its special onscreen keyboard, shown in Figure 8-5.

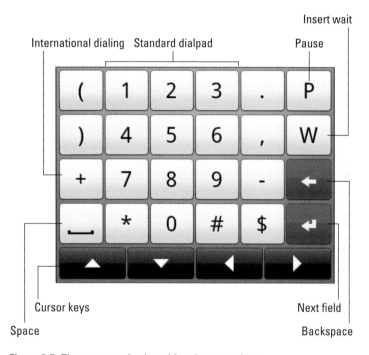

Figure 8-5: The onscreen keyboard for phone numbers.

Many keys on the dialpad keyboard are decorative, such as the parenthesis, hyphen, and space. Other keys have a special purpose, as illustrated in Figure 8-5.

The P and comma keys insert a brief pause into the dialing sequence. For example, in the number 91,5554275 the phone dials 91, pauses for two seconds, and then dials the rest of the number.

The W key inserts the Wait command into the dialing sequence. When the number is dialed, a pop-up message appears, prompting you to press the Yes button to dial the rest of the number.

Add a picture

The easy way to add a picture to a contact is to use a picture you already have, either a photo you've already taken that's stored in the phone or an image you've synchronized with the phone from your computer or an online picture-sharing website.

A more tortuous method is to meet the contact in person and convince them that you want to take their picture to add it to your phone's contact list. This effort requires a lot of social skills. Either way, the image ends up in the phone's Gallery. After it's there, follow these steps to add the picture to a contact:

1. **Locate and display the contact's information.**

2. **Press the Menu soft button and choose the Edit command.**

3. **Touch the Picture icon.**

   Refer to Figure 8-3.

4. **Choose the Gallery option.**

   Of, if the contact is right there with you, choose the Camera option and take a picture.

5. **Browse the Gallery to look for a suitable image.**

   See Chapter 15 for more information on using the Gallery.

6. **Touch the image you want to use for the contact.**

7. **Select the size and portion of the image you want to use for the contact.**

   Use Figure 8-6 as your guide. You can choose which portion of the image to use by moving the cropping box, and you can resize the cropping box to select more or less of the image, though the cropping box is always a square shape.

8. **Touch Save to assign the image to the contact.**

9. **Touch Save to complete the editing of the contact.**

Drag cropping box.

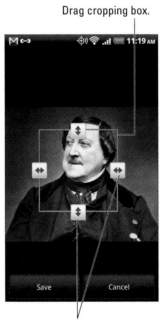

Resize cropping box.

Figure 8-6: Cropping a contact's image.

The image is now assigned, and it appears whenever the contact is refer-enced on your ThunderBolt.

To remove or change a contact's picture, follow Steps 1 through 3 and choose Remove Photo from the menu that pops up.

 You can add pictures to contacts on your Google account by using any com-puter. Just visit your Gmail Contacts list to edit a contact. You can then add to the contact any picture stored on your computer. The picture is eventually synced with the same contact on your ThunderBolt.

Set the primary phone number and email address

Rather than choose between multiple phone numbers and email addresses every time you summon a contact, you can specify which one you prefer to use most often. I'm happy to report that the ThunderBolt refers to your selec-tion as the *primary* phone number or *primary* email address, as opposed to using the dratted word *default*.

To set a person's primary phone number or email address, heed these steps:

1. **Display the contact's information.**

2. **Long-press the phone number or email address you want to use as the primary number or address.**

3. **Choose Set As Primary Number or Set As Default Address.**

Okay, so they slip in the word *default* when setting a primary email address. Damn them.

The word *Primary* appears by the main phone number or email address in a contact's address book entry.

Matching identical contacts

The phone can pull contacts from multiple sources (Facebook, Gmail, Twitter, its own memory), so you may discover duplicate contact entries in the People app. Rather than fuss over which entry to use, you can match up and link the contacts.

 Because the ThunderBolt is smart, the easiest way to link matching contacts is to take advantage of the Matched Contacts notification:

1. **Pull down the notifications and choose Matched Contacts Suggestions.**

   You see a list of suggested contact matches.

2. **Peruse the list.**

   You see a list of contacts stored on the phone or contacts found in your Google account. Beneath each item, you find potential matches with Facebook, Twitter, and other sources.

3. **Touch the link button by a contact to link the associated accounts.**

   Or press the Menu soft button and choose the Select All command to link all the matches at one time.

4. **Touch the OK button to link the accounts.**

You can also manually link a contact by touching the Link button in the upper right corner of the account's information screen, as shown earlier, in Figure 8-2. After touching the Link button, you see either a list of linked accounts or potential links you can make.

Making a contact group

The ThunderBolt automatically synchronizes any contact groups you may have in Gmail or from other sources. You can use the groups to send a bunch of people email or text messages. It's handy.

Review your groups by starting the People app. Choose Groups from the bottom, as shown earlier, in Figure 8-1. You see all groups available in your phone. Touch a group to view its members, as shown in Figure 8-7.

To add members to a group, touch the Add Contacts to Group button, shown in Figure 8-7. Touch the gray square by a contact to place a green check mark there and add the contact to the group. Touch the Save button to add the contact. (Press the Back soft button to dismiss the onscreen keyboard to see the Save button.)

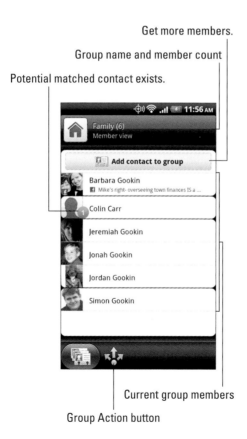

Get more members.

Group name and member count

Potential matched contact exists.

Current group members

Group Action button

Figure 8-7: Members of a group.

Use the Group Action button (refer to Figure 8-7) to send everyone in the group a text message or an email.

You can create new groups at any time. Follow these steps:

1. **Start the People app.**

2. **Touch the Groups button.**

3. **Touch the Add Group button.**

4. **Type a name for the group.**

   Keep the group name simple and descriptive. For example, I have a group named Cohorts, which contains people with whom I plot the overthrow of our city hall.

5. **Optionally, touch the camera button to assign an image to the group.**

6. **Touch the Add Contacts to Group button.**

7. **Touch the box by a contact's name to place a green check mark there and add the contact to the group.**

8. **Touch the Save button to add the group's members.**

   You may need to press the Back soft button to dismiss the onscreen keyboard so that you can see the Save button.

9. **Touch the Save button to create the group.**

   The group is added to the list.

You can now access the group's Send menu, shown in Figure 8-7, to send the group an email or a text message.

- To remove a group, long-press the group name and choose the command Delete Group from the pop-up menu. Touch the Yes button to confirm.

- The members of the Favorites group appear in the Favorites widget on the Home screen.

- The Frequent group lists folks you frequently contact. It's automatically created and maintained.

- When adding members to a group, it helps immensely to use the Search Contact text box to help quickly narrow the list.

- The best way to create Gmail contact groups is by using Gmail on the Internet. Go to http://gmail.google.com.

- See Chapter 10 for more information on sending email with your ThunderBolt; Chapter 9 covers text messaging.

Sharing a contact

Just as you can import contacts into the phone, you can send a contact out of the phone. It's a way you can share the people you love.

The ThunderBolt can convert any address book entry into the common vCard file format. *vCards* are used by computer email programs to store contact information, which makes that file format popular for sharing information about your people.

Follow these steps to share a contact:

1. **Open the People app.**

2. **Long-press the person you want to share.**

3. **Choose the command Send Contact As vCard from the pop-up menu.**

4. **Choose the details you want to send.**

    Add or remove green check marks by the items you want to include in the vCard.

5. **Touch the Send button.**

6. **Compose and send the email message.**

At this point, the rest of the operation works like sending an email message, which is covered in Chapter 10. The contact's information is saved as a vCard file, which is attached to the message.

Getting rid of a contact

Every so often, consider reviewing your contacts. Purge those folks whom you no longer recognize or you've forgotten. It's simple:

1. **Locate the contact in the People app and display the contact's information.**

2. **Press the Menu soft button.**

3. **Choose the Delete command.**

4. **Touch OK to remove the contact from your ThunderBolt.**

If you delete a Google contact, it's also removed from your Google account on the Internet.

For some linked accounts, such as Facebook, deleting the account from your phone doesn't remove the human from your Facebook account. A warning might appear to explain as much.

Removing a contact doesn't kill the person in real life.

Part III
Beyond the Basic Phone

The 5th Wave   By Rich Tennant

Cell Phones

"This model comes with a particularly useful function — a simulated Static button for breaking out of long-winded conversations."

In this part . . .

There are several criteria for being an ancient Greek god. Beyond immortality are other god-like abilities, among them transmogrification, tele-portation, and a supercharged libido. A god also needs the ambition to blithely kill his parents, sib-lings, children, aunts, and uncles, and even the Schwan's delivery guy. Combine all that with the temperament and emotional maturity of a junior high school brat and you have enough material on a curriculum vitae to qualify for deification.

My point about the ancient Greed gods is that their abilities weren't limited to simply being immortal. They had to do other things, and specialize in different areas, in order to earn their places in the pantheon. Similarly, your ThunderBolt phone has communications abilities far beyond just being a phone. Though these abili-ties may not elevate the phone to god status, they're impressive, and the topic of this part of the book.

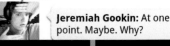

The Texting Craze

*T*he most popular way to communicate on a cell phone isn't by talking. It isn't by using email. It isn't by using the Internet. No, the most popular way to communicate on a cell phone is to send a text message. That's it: a brief, crudely typed, to-the-point message. It's a *text* message because that's all it is: just text. The process is known as *texting*.

I don't need to explain to anyone over the age of 25 that texting is an obsession. It's the primary way that most teenagers communicate. Shorter than an email, faster than a phone call, a text message is a handy, swift way to stay in touch.

⌐ thank Michelle for the
ely cake. It's not often that

U Hav a Msg

The place to visit for your texting pleasure on the ThunderBolt is the app named Messages. It's where you read your texting conversations, start new ones, and generally sate your voracious text messaging appetite.

⌐ Some Android applications can affect messaging. You're alerted to whether a program affects messaging before it's installed. See Chapter 18.

✔ Your cellular plan may include a preset limit for text messages — for example, 200 messages per month — a laughably unrealistic number. If you exceed this number, you're charged a per-message fee. You can sign up for unlimited texting plans, well suited for the parents of teenagers.

✔ By the way, the charge for messaging is per message whether it's sent or received. Yikes.

✔ The nerdy term for texting is *SMS,* which stands for Short Message Service.

Using the Messages app

The Messages app comes prepasted to the main Home screen, or you can pluck it from the assortment of apps found on the All Apps screen. Figure 9-1 illustrates the main screen for the Messages app, which lists all ongoing conversations.

Ongoing conversations

Touch to view the conversation.

Unread message waiting

Send a new message.

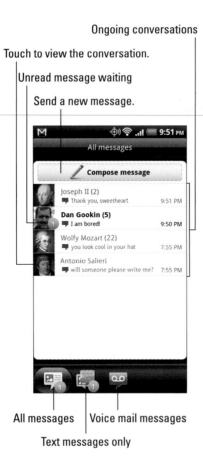

All messages   Voice mail messages

Text messages only

Figure 9-1: The Messages app.

Unviewed messages are highlighted, as shown in Figure 9-1. Simply touch an item to review the conversation or to start a new conversation, as covered in the next section.

- ✔ The tabs at the bottom of the Messages app window let you view text messages as well as voice mail messages. The All Messages screen lists both text and voice mail messages.

- ✔ The voice mail messages accessed from the Messages app are either carrier voice mail or Visual Voicemail. Google Voice messages don't appear in the Messages app.

- ✔ The Voicemail app is a gateway into the Messages app, with the Voicemail messages tab viewed first.

- ✔ See Chapter 7 for more information on voice mail.

Composing a new text message

You can send a text message to just about anyone who has a cell phone. It works like this:

1. **Open the Messages app.**

2. **Touch the Compose Message button.**

   A message composition window appears, which also tracks your text conversation, similar to the one shown in Figure 9-2.

3. **Input a mobile number or a contact name in the To field.**

   If the text you type matches one or more existing contacts, you see those contacts displayed. Choose one to send a message to that person; otherwise, continue typing the phone number or email address.

4. **Touch the Add Text box.**

5. **Type your text message.**

   You have only 160 characters to make your point. A handy counter shows you how many characters you have left.

6. **Touch the Send button to send the message.**

The message is sent instantly, or as quickly as the cellular network can send it. You can wait for a reply or do something else with the phone, such as snooze it or choose to talk with a real person, face-to-face. Or, you can always get back to work.

See the later section "Receiving a text message" for information on replies to your messages.

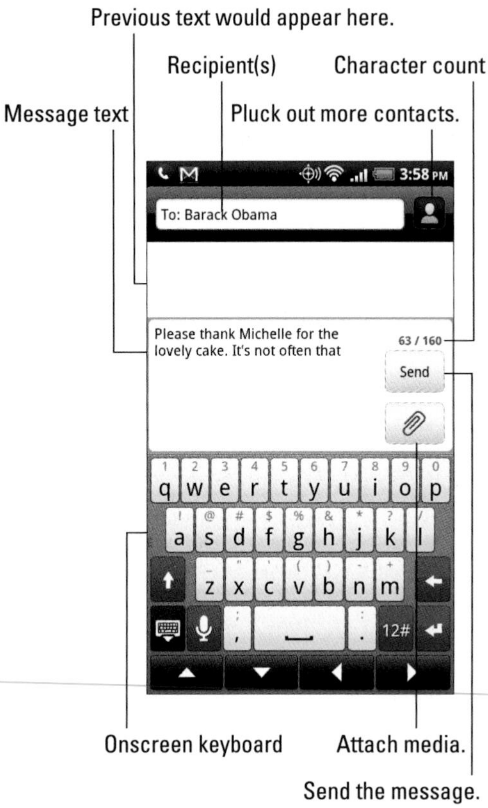

Previous text would appear here.

Recipient(s)   Character count

Message text   Pluck out more contacts.

To: Barack Obama

Please thank Michelle for the
lovely cake. It's not often that

63 / 160

Send

Onscreen keyboard   Attach media.

Send the message.

Figure 9-2: Typing a text message.

> ✒ Add a subject to your message by touching the Menu soft button and choosing More, and then Add Subject. The subject appears only on your phone, however. It's used only to identify separate message threads with the same contact.

> ✒ On Android devices, any smiley faces you put in a text message appear as special Android icons. Press the Menu soft button and choose More and then Insert Smiley to see the gamut.

> ✒ Phone numbers and email addresses sent in text messages become links. You can touch a link to call its number or visit its web page.

> ✒ Yes, you can send a single text message to multiple people. See the side-bar "Which is better: text message or email message?" for my thoughts on sending a text message versus sending an email.

> ✒ To cancel a message, press the Menu soft button and choose the Discard command. Touch the OK button to confirm.

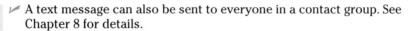
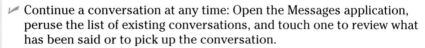

✏ A text message can also be sent to everyone in a contact group. See Chapter 8 for details.

✏ Continue a conversation at any time: Open the Messages application, peruse the list of existing conversations, and touch one to review what has been said or to pick up the conversation.

✏ Do not text and drive. Do not text and drive. Do not text and drive.

Sending a text message to a contact

You can send a text message to anyone in your address book, if they have a cell phone or another mobile gizmo that can pick up text messages. Here's how it works:

1. **Open the People app.**

2. **Choose a contact to whom you want to send a text message.**

3. **Touch the Send Message item in the Action area.**

   You can send a text message only to a cell phone, and not every cell phone has the ability to receive text messages.

4. **Type the message text.**

5. **Touch the Send button.**

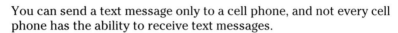

You can create a direct message shortcut for a contact you text often. Place the shortcut on the Home screen for quick messaging action. See Chapter 8 for details.

Which is better: text message or email message?

Sending a text message is similar to sending an email message. Both involve sending a message instantly to someone else. Yet both methods of communication have their advantages and disadvantages.

Text messages are short and to the point. They're informal, more like quick chats. Indeed, the speed of reply is often what makes text messaging useful. But, like sending an email, sending a text message doesn't guarantee a reply.

An email message can be longer than a text message. You can receive email on any computer or device, such as your ThunderBolt, that can access the Internet. Email message attachments (pictures, documents) are handled better, and more consistently, than text message (MMS) media. You can also reply to everyone in an email message, whereas you can send multiple recipients only a single text message.

Finally, though email isn't considered formal communication, not like a paper letter or a phone call, it ranks higher in importance than a text message.

Popular text message abbreviations

I can't decide whether texting isn't about using proper English or whether the 160-character limitation on a text message forces people to spell conveniently as opposed to properly. Either way, the texting craze has introduced a host of abbreviations and word shortcuts into the English language. Terms such as *LOL* and *BRB* have become part of the popular culture.

The texting acronyms find their roots in the Internet chat rooms of yesteryear. Regardless of their source, you might find them handy for typing messages quickly. Or, maybe you can use this reference for deciphering an acronym's meaning (you can type acronyms in either upper- or lowercase):

<3	Heart, love	K	Okay	TC	Take care
2	To, also	L8R	Later	THX	Thanks
411	Information	LMAO	Laughing my [rear] off	TIA	Thanks in advance
BRB	Be right back			TMI	Too much information
BTW	By the way	LMK	Let me know	TTFN	Ta-ta for now (goodbye)
CYA	See you	LOL	Laugh out loud		
FWIW	For what it's worth	NC	No comment	TTYL	Talk to you later
FYI	For your information	NP	No problem	TY	Thank you
GB	Goodbye	OMG	Oh my goodness!	U2	You, too
GJ	Good job	PIR	People in room (watching)	UR	Your, you are
GR8	Great			VM	Voice mail
GTG	Got to go	POS	Person over shoulder (watching)	W8	Wait
HOAS	Hold on a second			XOXO	Hugs and kisses
IC	I see	QT	Cutie	Y	Why?
IDK	I don't know	ROFL	Rolling on the floor, laughing	YW	You're welcome
IMO	In my opinion	SOS	Someone over shoulder (watching)	ZZZ	Sleeping
JK	Just kidding				

Receiving a text message

Whenever a new text message comes in, you see a message appear at the top of the touchscreen. The message goes away quickly, and then you see the New Text Message notification, shown in the margin.

Opting out of text messaging

You don't have to be a part of the text messaging craze. Indeed, it's entirely possible to opt out of text messaging altogether. Simply contact your cellular provider and tell them that you want to disable text messaging on your phone. They will happily comply, and you'll never be able to send or receive a text message again.

People opt out of text messaging for a number of reasons. A big one is cost: If the kids keep running up the text messaging bill, it's often easier to simply disable the feature than to keep paying all the usage surcharges. Another reason is security: Though it's uncommon, viruses and spam have been sent via text message. If you opt out, you don't have to worry about receiving unwanted text messages.

To view the message, pull down the notifications, as described in Chapter 3. Touch the messaging notification and that conversation window immediately opens.

When a new message floats in, the ThunderBolt plays a notification ringtone. The ringtone can be changed, as discussed in Chapter 6.

Forwarding a text message

It's possible to forward a text message, but it's not the same as forwarding email. In fact, when it comes to forwarding information, email has text messaging beat by about 160 characters.

The bottom line is that you can forward only the information in a text messaging cartoon bubble, not the entire conversation. Here's how it works:

1. **If necessary, open a conversation in the Messages app.**

2. **Press the text entry you want to forward.**

3. **From the menu that appears, choose the Forward command.**

   From this point on, forwarding the message works like sending a new message from scratch:

4. **Type the recipient's name (if the person is a contact) or type a phone number.**

   The text you're forwarding appears already written in the text field.

5. **Touch the Send button to forward the message.**

If you'd rather forward the text message in a new email, choose the Select Text command in Step 3. Select and copy the message's text per the directions found in Chapter 4. Open the Email program and paste the message text into a new email message, and then go on from there.

- Refer to Chapter 10 for more information on composing email on your ThunderBolt.

- When it comes to forwarding multimedia attachments, I recommend that you first save the attachment to your phone. After it's been saved, you can use either the Gmail or Email app to attach the media, as described in Chapter 10.

Text Messages with Pictures and Stuff

Text messages don't have to be text-only. You can attach media to a text message, such as video, a picture, or an audio tidbit. By doing so, you transform the simple text message into something new, into a multimedia message.

- A text message with a multimedia attachment ceases to be a text message and becomes an MMS, which stands for Multimedia Messaging Service.

- There's no need to run a separate program or do anything fancy to send media in a text message; the same Messages app is used on the ThunderBolt for sending both text and media messages. Just follow the advice in this section.

- Not every mobile device has the ability to receive MMS messages. Rather than receive the media, the recipient is directed to a web page where the media can be viewed on the Internet.

Attaching media to a message

The key to adding media — video, audio, pictures — to a multimedia message is to use the Attach button, shown earlier, in Figure 9-2. Touching the button displays the Attach menu, from which you can pluck a variety of options to virtually paperclip to your text message. The general procedure works like this:

1. **Compose a text message as you normally do.**

   Fill in the To field, or simply continue an existing conversation.

2. **Touch the Attach button.**

   Refer to Figure 9-2 for the location of the Attach button.

After touching the Attach button, you see a pop-up menu listing various media items you can attach to a text message. Here's a summary:

Picture: Choose an image stored in the Gallery.

Video: Choose a video you've taken with the phone and stored in the Gallery.

Audio: Browse the folders on the ThunderBolt to find an audio file to attach.

App Recommendation: Send a link to the Amazon Marketplace, which the recipient can touch to download an app.

Location: Send your latitude and longitude, address, or a Google Maps web link.

Contact: Send information about one of your contacts as a vCard attachment.

Appointment: Send information from the Calendar app in the form of a vCalendar attachment.

Slide show: Create a series of images chained together, which you can send in one batch.

More options may appear on the menu, depending on which apps you have installed on your ThunderBolt.

3. **Type a message to go with the media attachment.**

   The attachment appears in the message as a large button, shown in Figure 9-3.

4. **Touch the Send button to send your media text message.**

Figure 9-3: Text message attachment.

In just a few, short, cellular moments, the receiving party will enjoy your multimedia text message.

✔ Not every mobile device is capable of receiving multimedia messages.

✔ Be aware of the size limit on the amount of media you can send; try to keep your video and audio attachments brief.

✔ You may be prompted to shrink or compress larger images or videos.

✔ Choosing some options displays an additional menu, from which you choose which app to complete the action. For example, when attaching audio, you get to choose between the Voice Recorder app for recording a message or the Music app for sending an audio clip.

✔ Options exist in the Camera app for taking smaller-size pictures and videos. See Chapter 14 for more information.

Receiving a multimedia message

Multimedia attachments come into your ThunderBolt just like any other text message does, but you see a thumbnail preview of whatever media was sent. Touch the attachment to preview its contents. Pictures and videos show up full-screen.

To save a picture attachment in a multimedia message, follow these steps:

1. **Touch the picture attachment.**

   The attachment appears on a screen by itself.

2. **Touch the screen to summon a menu.**

3. **Touch the Save icon.**

   The picture (or pictures) in the attachment appear in a list.

4. **Touch the gray check box by the pictures you want to save.**

   A green check mark appears in the box, marking the image.

5. **Touch the Done button.**

   The image is saved.

You can use the Gallery app to view the saved image and video attachments. They're found in the Downloads folder. See Chapter 15.

Some types of attachments, such as audio, cannot be saved.

To return to the message after playing its media attachment, press the Back soft button.

Clean Up Your Messages

The topic here isn't propriety or using foul language in your text messages. No, the topic is message management.

When you exchange messages with someone, you create a *conversation*. To remove that conversation, and erase all the text that's been exchanged, follow these exact steps:

1. **Go to the main screen in the Messages app.**

   It's the screen that lists all your conversations, as shown earlier, in Figure 9-1.

2. **Long-press the message thread you want to delete.**

3. **Choose Delete from the pop-up menu.**

4. **Touch the OK button to confirm.**

   The conversation is gone.

When you've been particularly sloppy, or you merely want to remove multiple conversations, follow these steps:

1. **Touch the Menu soft button.**

2. **Choose Delete.**

3. **Touch the box next to every conversation you want to zap.**

   Obviously, if you want to keep one, don't touch its box.

   A red X appears by conversations slated for execution.

4. **Touch the Delete button.**

The selected message threads are gone.

Other Messaging Options

Texting has its limitations. There's the 160-character limit, but more importantly, your cellular account may have a maximum limit on the number of text messages you can send *and* receive during the billing cycle. You don't have to go suddenly silent when you hit the maximum limit, or risk running up a horrendous bill. That's because you can take advantage of some texting and chatting alternatives on your ThunderBolt, as covered in this section.

Messaging instantly with Mobile IM

In the land of two-letter computer acronyms, IM stands for *instant messaging*. It was a craze for a while, and I suppose it's still popular on Facebook and other web pages. On your ThunderBolt, the app Mobile IM allows you to bring in some of your Internet instant messaging accounts so that you can enjoying chatting with your IM buddies on your phone.

Instant messaging works best when you've already set up an account on an IM service, such as Windows Live, Yahoo!, or AOL IM (AIM). Further, you should also have an account with Verizon: Visit `www.verizonwireless.com` and click the My Verizon button to set up an account, if necessary.

After you do that, visit the All Apps screen on your ThunderBolt and run the Mobile IM app.

The first time you open Mobile IM, you're prompted to choose an IM service. Log in to the service using your email address or account name and password.

After you sign in to your accounts, you see any instant messages you've received appear on the screen, similar to the way text messages appear in the Messages app.

- ✏ To add more accounts to the Mobile IM app, touch the Add Account button.

- ✏ Press the Menu soft button and choose the Accounts command to see all your IM accounts.

- ✏ Unless you've chosen the option to stay signed in (Auto Sign-In), quitting the Mobile IM app logs you out of your IM network.

Talking with Google Talk

Another way to chat it up with your online friends is the Google Talk service. This Internet-based chatting service can be found on the main Gmail page. In fact, you already have a Google Talk account if you have a Google account.

On your ThunderBolt, you can use the Talk app to connect you with Google Talk. It's not really a texting app, but, rather, a chat app. You can summon a list of friends, all configured from your Google account, and chat it up — if they're available online for vicious typing.

As with other Internet-based services, my advice is to configure Google Talk on your computer first, and then you'll find the same friends available on your ThunderBolt.

Email Headquarters

A long time ago, writing a letter was an art. You wrote in cursive, and to great effect. Letters were personal, intelligent, often poetic. Protocols existed for writing to friends, businesses, or important people. There was even this stuff called *stationery*. That's all gone now, replaced by the instant convenience and digital ambiguity of electronic mail, or email.

The ThunderBolt makes sending and receiving email convenient because wherever you take the phone, your email comes with you. The phone receives email messages from all your accounts, letting you read the missives from wherever you are, as they come in. You can also use the phone to compose new messages. This chapter explains how it works.

The Big Email Picture

The ThunderBolt features two email apps: one for Google's Gmail and another one, Mail, that can handle other types of email.

The Gmail app hooks directly into your Gmail account. It's basically an echo of your Gmail account on the Internet.

The Mail app is used to connect to a variety of email accounts and types, such as email from your ISP, your business or organization, plus web-based email such as AOL, Hotmail, Microsoft Live Mail, Yahoo!, and whatever other email accounts you have.

Beyond the different apps, your messages are handled much the same as they are on your computer: Using your ThunderBolt, you can receive email, compose new messages, forward email, work with contacts, attach media, and perform all your other email activities. As long as the phone has a data connection, email works just fine.

- ✔ You can run the Gmail and Mail apps by touching the All Apps button on the Home screen and then locating the apps on the All Apps screen.

- ✔ The Mail app icon appears on the main Home screen. The Gmail app can be found on the second Home screen to the left of the main Home screen. You can move those icons around, if you like; directions are offered in Chapter 22.

- ✔ A Gmail account was created for you when you signed up for a Google account. See Chapter 2 for more information about setting up a Google account.

- ✔ Though you can use your phone's web browser (the Internet app) to visit the Gmail website, you should use the Gmail app to pick up your Gmail.

- ✔ Likewise, you should use the Mail app to pick up your email from AOL Mail, Hotmail, Yahoo! Mail, or another web-based email system.

- ✔ If you forget your Gmail password, visit this web address:

      www.google.com/accounts/ForgotPasswd

- ✔ Every so often, Google updates the Gmail app, adding new features. If anything changes after this book goes to press, refer to my website for updates and additional information:

      www.wambooli.com/help/phone

Email Setup and Configuration

Gmail is automatically set up for use on the ThunderBolt. Other email accounts force you to go through some gyrations. The process is similar to configuring an email program on your computer, though you're doing it with a phone and a touchscreen keyboard. This difference makes the process extra fun.

Adding an email account

To access your non-Gmail email, you need to set up and configure your non-Gmail email accounts on the ThunderBolt. This operation is handled by the Mail app, which is handy because you use it anyway to retrieve your non-Gmail email.

Non-Gmail email accounts include any free web-based email, your ISP's email, or maybe email from your company or another larger, intimidating organization.

Obey these steps to set up your non-Gmail email:

1. **Open the Mail app.**

   You should see the Set Up Email screen, listing the various types of email accounts you can add (though your selection is by no means limited to these icons). If you see the screen, skip to Step 4; otherwise, you probably already have some accounts configured. Continue with Step 2:

2. **Press the Menu soft button.**

3. **Choose More and then choose New Account.**

   The New Account command might be visible without having to first choose the More command.

4. **Choose an icon to add an email account from that service or provider.**

   For example, choose AOL or Yahoo! if you have an email account at AOL or Yahoo!.

   Choose Win Live Hotmail if you have an account on Hotmail, MSN, or Windows Live.

   If you have an ISP account, such as a Comcast or Road Runner account, choose Other (POP3/IMAP). For work, you probably have to choose Microsoft Exchange ActiveSync.

   Over the next few steps, you need to know some information about your email account. If you don't have this information, threaten someone who does.

5. **Type the email address you use for the account.**

   Type a full email address.

6. **Type the password for the account.**

   After you add the first account, you see an option to make the new account (the one you're adding) the main account. If that's what you want to do, choose one of the three options in Step 7:

7a. *To configure a web-based email account,* **such as Hotmail or AOL, touch the Next button; skip to Step 13.**

7b. *To configure an Exchange Server account,* **fill in the information on the Exchange Server Settings screen.**

   Use the information provided by your organization to fill in all appropriate fields; skip to Step 13.

7c. *To configure an ISP email account,* **choose Manual Setup.**

   Manual setup is required for non-web-based email because you need to supply specific information to configure it. That's the tedious part.

8. **Because you're configuring your ISP email account, touch the POP3 Account button.**

9. **Fill in the Incoming Settings information.**

   For example, for a POP3 server, you input the proper POP3 server name provided by your ISP. The ThunderBolt guesses at the name, though that name (in the POP3 Server field) might be incorrect.

   The other fields on the Incoming Settings screen are probably all okay; don't mess with them.

10. **Touch the Next button.**

11. **Fill in the Outgoing Server Settings screen.**

   Though most of the information on the screen should be okay, check the SMTP Server field to ensure that it shows the server name as provided by your ISP.

   If the server doesn't require verification, remove the check mark by the Login Required box.

12. **Touch the Next button.**

   The Email Account screen appears.

13. **Type a name for the account in the Account Name field.**

   I use the name of the service for the account name, such as Yahoo or Windows Live. For my ISP's email account, I gave it the name *Main* because it's my primary email account.

14. **Optionally, change the name in the Your Name field.**

    It's the name that appears for outgoing mail. For my email account, I changed the name from my email address to **Dan Gookin**, which happens to be my name in real life.

15. **If the option is available and you want to make the account your main email account, place a check mark by Make This My Default Mail Account.**

    Only one account can be the main account. I recommend using your ISP email account, if you have one set up.

16. **Touch the Finish Setup button.**

    Your email account is set up. The ThunderBolt does a quick check for pending mail, and then you see the Inbox for that email account.

You can set up all your email accounts on your ThunderBolt — even multiple accounts for different services. So, if you have more than one Microsoft Live account, for example, add them all.

Not every web-based email account can be accessed by the ThunderBolt. When doubt exists, you see an appropriate warning message. In most cases, the warning message also explains how to properly configure the web-based email account to work with your ThunderBolt.

Changing the default email account

The ThunderBolt sends your email using whichever account you're accessing at the time — unless you're viewing the All Accounts inbox when you choose to compose a message. In that case, the default account is used to send the message. To set or change the default account, obey these steps:

1. **Open the Mail app.**

2. **Press the Menu soft button and choose Accounts List.**

    You see a list of all the email accounts you've set up on your phone, similar to the ones shown in Figure 10-1.

3. **Choose the account you want to make the new default account.**

    You're taken to that account's inbox, but you're not done yet:

4. **Press the Menu soft button and choose More and then Settings.**

5. **Choose General Settings.**

6. **Place a green check mark in the box by Set As Default Account.**

7. **Press the Back button twice to return to the account's inbox.**

Universal inbox   Unread messages

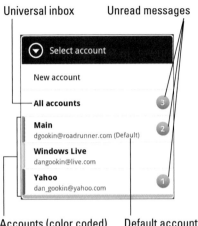

Accounts (color coded)   Default account

Figure 10-1: Email accounts in the Mail app.

Only one account can be the default.

Making a signature

Email signature shouldn't be optional on a cell phone. It's just nice to let people know — in case of grievous typos — that the messages they're receiving were composed on a mobile device. As an example, here's my email signature:

```
DAN

This was sent from my ThunderBolt.
Please forgive the typos.
```

To create a signature for Gmail, obey these directions:

1. **Start Gmail.**

2. **Press the Menu soft button and choose More and then Settings.**

   If you see no settings, choose Go to Inbox and repeat this step.

3. **Choose Signature.**

4. **Type or dictate your signature.**

   The signature can be on two lines, even though it looks like just one line is available.

5. **Touch OK.**

6. **Press the Back soft button to return to your Gmail inbox.**

You can obey these same steps to change your signature; the existing signature shows up after Step 4.

To set a signature for the Mail app, follow these steps:

1. **Press the Menu soft button.**

2. **Choose Accounts List.**

3. **Choose an account.**

   Even though the default account does all the sending, each individual account has its own email signature.

4. **Press the Menu soft button and choose More and then Settings.**

5. **Choose General Settings.**

6. **Choose Signature.**

7. **Add or edit your email account signature.**

   You may see a preset signature, one that may plug your wireless provider. Whatever. Feel free to type whatever you like as a signature.

8. **Touch the Save button.**

9. **Ensure that there's a check mark by the item Use Signature.**

10. **Press the Back soft button twice to return to the account's inbox.**

The signature you set appears in all outgoing messages for the account you chose in Step 3. To set the signature for another account, repeat these steps for that specific account.

You can copy and paste your signature from one account to another. See Chapter 4 for information on how copy-and-paste works on the ThunderBolt.

Configuring email options

The Mail app offers plenty of options and settings. Rather than bore you by listing them all, I thought I'd mention what I feel are the two most important: Email check frequency and leaving mail on the server. Follow these steps in the Email app:

1. **Press the Menu soft button and choose Accounts List.**

2. **Select the account you want to configure.**

   Each account must be configured separately.

3. **Press the Menu soft button and choose More and then Settings.**

4. **Choose Send & Receive.**

5. **Ensure that there's no mark by the item Delete Mail on Server.**

   By keeping the mail on the server, you allow your computer to get the same messages later. Otherwise, email you receive on your phone won't be available on your computer.

6. **Choose Update Schedule.**

7. **Set the peak time when you want the ThunderBolt to vigorously check your messages.**

   Choose Days to set which days; choose Peak Time Start and then Peak Time End to set when the phone will start and stop actively checking for new messages, respectively.

8. **Choose Peak Times to set the update frequency.**

   The default value is 15 minutes. Because I get a lot of email, I changed this setting to Every 5 Minutes.

9. **Press the Back soft button thrice to return to the account's inbox.**

Repeat these steps to configure each of your email accounts.

- ✐ If you elect to delete mail on the server (refer to Step 5), any mail you receive on the phone isn't received by any other device, including your computer.

- ✐ Gmail is checked all the time for new messages, so there's no need to make an update-frequency setting in that app. Also, Gmail messages are kept on the server as well as on the phone; they're not deleted unless you choose to delete a message.

You've Got Email

You're alerted to the arrival of a new email message in your ThunderBolt by a notification icon as well as by a ringtone or another sound. The notification icon differs between a new Gmail message and a Mail message.

 For a new Gmail message, you see the New Gmail notification, shown in the margin, appear at the top of the touchscreen.

For a new email message, you see the New Mail notification.

To deal with the new-message notification, pull down the notifications and choose the appropriate one. You're taken directly to the appropriate inbox to read the new message.

Refer to Chapter 3 for information on accessing notifications.

Checking the inbox

To peruse the mail you have, start your email program — Gmail for your Google mail or Mail for other mail you have configured to work with the ThunderBolt — and open your electronic inbox.

To check your Gmail inbox, start the Gmail app. It's found on the Applications screen. The Gmail inbox is shown in Figure 10-2.

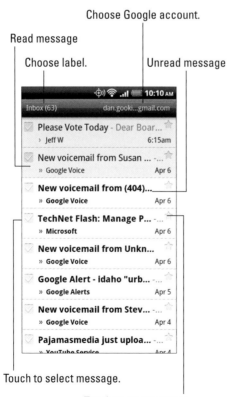

Choose Google account.

Read message

Choose label.   Unread message

Touch to select message.

Touch to star message.

Figure 10-2: The Gmail inbox.

To get to the inbox screen when you're reading a message, press the Back soft button. Or, you can press the Menu soft button and choose the command Go to Inbox.

To check your non-Gmail email, you open the Mail app. Because this app can grab its mail from multiple accounts, you see multiple inboxes. To see all your new email in one spot, you can access the All Accounts universal inbox, shown in Figure 10-3. Then ensure that the Inbox item is chosen from the tab at the bottom of the screen.

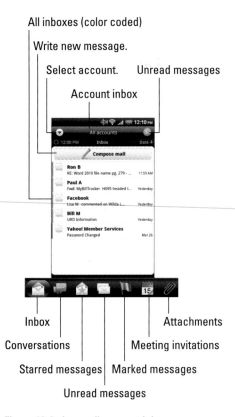

All inboxes (color coded)

Write new message.

Select account.   Unread messages

Account inbox

Inbox   Attachments

Conversations   Meeting invitations

Starred messages | Marked messages

Unread messages

Figure 10-3: An email account inbox.

To view your universal inbox, which shows all incoming email for all your email accounts, touch the Select Account button (shown in Figure 10-3) and choose All Accounts. Likewise, to view messages in a specific account, choose the account name from the same menu.

Gmail is organized using *labels,* not folders. To see your Gmail labels from the inbox, touch the Menu soft button and choose Go to Labels.

Reading a message

To read a message, choose it from either the Gmail or Mail inbox, described earlier in this chapter. The message appears on the screen, as shown in Figures 10-4 and 10-5 for Gmail and Mail, respectively.

From (account)

Label(s)   Show reply toolbar.

Message subject   Reply

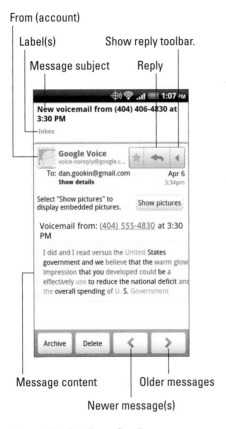

Message content   Older messages

Newer message(s)

Figure 10-4: Reading a Gmail message.

Touch the Reply toolbar button (see Figure 10-4) to see the Reply All and Forward buttons.

Sender contact info

Message subject

Newer messages Older messages

Message content

Figure 10-5: Reading a Mail message.

Reading a message and working with a message operate much the same in any email program you've used. Refer to the nearby figures for locating common email actions, as described in this list:

✔ To reply to a message, touch the Reply button in the Gmail app or in the Mail app.

✔ The Reply All command is accessed by touching the Reply toolbar in Gmail. In the Mail app, choose Reply All from the button on the screen.

✔ To forward a Gmail message, use the Reply toolbar to find the Forward button. In the Mail app, press the Menu soft button and choose More and then Forward.

✔ Refer to the later section "Compose an Electronic Epistle" for information on (surprisingly) writing a new electronic message, which also applies when you forward or reply to an email.

✔ To delete a message in Gmail, touch the Delete button. In the Mail app, press the Menu soft button and choose the Delete command.

✔ Use the Older and Newer buttons to continue to read your email messages.

✔ When you're done with a message, you can press the Back soft button to return to the inbox.

✔ Use the Reply All command only when everyone else *must* get a copy of your reply. Because most people find endless Reply All email threads annoying, use the Reply All option judiciously.

Saving an email attachment

Receiving an email attachment on your phone works similarly to receiving an attachment on your computer. The main difference is that you can't work with files on your ThunderBolt as well as you can on a computer: You cannot edit a Word document, for example. But that's okay because you can still access the attachments.

Email messages with attachments are flagged in the inbox with paper clip icons, which seems to be the standard I-have-an-attachment icon for most email programs. How the attachment shows up depends on whether you're using the Gmail app or the Mail app.

In Gmail, pictures may show up automatically, in line with the rest of the message. You may see a Preview button, which is used to view the image on another screen, and a Download button, which is used to save the image to the ThunderBolt's MicroSD card. You can then view the image by using the Gallery app.

Most Gmail attachments, however, show up as shown in Figure 10-6. The filename appears next to the Preview button. Touch the button to see (or hear) the attachment.

Figure 10-6: An attachment in a Gmail message.

In the Mail program, attachments are listed at the top of the message, as shown in Figure 10-7. Touch the Show More arrow to peruse the attachments. Touch the attachment, shown in the figure, to display or play the message.

Touch to view or play.

Show/Hide attachments.

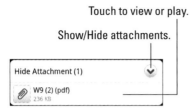

Figure 10-7: An attachment in a Mail message.

In cases with no obvious Save button, viewing the attachment also down-loads it to your phone. You can find most picture attachments by using the Gallery app and looking in the Downloads folder. Audio apps are found in whichever audio player you used to listen to the attachment, such as the Music app.

- ✔ You may be asked to choose which apps to use for previewing the attachment. The choices presented on the Complete Action Using menu depend on the apps installed on your phone.

- ✔ Some attachments can be neither previewed nor saved. In these cases, use a computer to fetch the message and attempt to open its attach-ment. Or, you can reply to the message and inform the sender that you cannot open the attachment on your phone.

- ✔ Sometimes, pictures included in an email message aren't displayed. You find the Show Pictures button in the message, which you can choose to display the pictures; refer to Figure 10-4.

- ✔ Any email attachments you save are stored on the MicroSD card, in the `Download` folder. The ThunderBolt doesn't come with file management apps, though you can download them from the Android Market. Or, you can manage the MicroSD card's files when the ThunderBolt is mounted on your computer's storage system. See Chapter 20 for information.

Searching email

You can use the Search soft button to search the email on your ThunderBolt. This feature works best with Gmail; you can search only the message headers in the Mail app.

To search for any tidbit of text in Gmail, heed these directions:

1. **Open the Gmail inbox.**

2. **Touch the Search soft button.**

3. **Type the text to find.**

   You can also dictate the text by first pressing the Microphone button on the keyboard and then speaking what you're trying to find.

4. **Touch the Search button to begin the search.**

   Peruse the results.

To perform a wider search throughout the entire ThunderBolt, visit the Home screen and touch the Search soft button to summon the Search Anywhere app.

Compose an Electronic Epistle

Because people love getting email, why not make someone's day and send them a message? Email is quick and easy to compose on your ThunderBolt, using either the Gmail or Mail app. Be clever. Be pithy. Be sure to read this section.

Writing a new Gmail message

New Gmail messages are created on your phone similarly to the way they're created on your computer. That way, if you're frustrated by using Gmail on your computer, you'll enjoy the same frustration on your phone.

Figure 10-8 shows the basic Gmail composition setup.

Follow these steps to craft your missive:

1. **Start the Gmail app.**

2. **Ensure that you're viewing the inbox.**

   If not, press the Back soft button.

3. **Press the Menu soft button.**

4. **Choose Compose.**

   A new message screen appears, looking similar to Figure 10-5 but with none of the fields filled in.

Message content

Message subject   Send message.

To (recipient)   Save as draft.

Onscreen keyboard

Figure 10-8: Writing a new Gmail message.

5. **Type the first few letters of a contact name, and then choose a matching contact from the list that's displayed.**

   You can also send to any valid email address not found in your Contacts list, by typing that address.

   To summon the CC field, press the Menu soft button and choose the command Add Cc/Bcc.

6. **Type a subject.**

7. **Type or dictate the message.**

8. **Touch the Send button to whisk your missive to the Internet for immediate delivery.**

Copies of the messages you send are saved in your Gmail account, which is accessed from your ThunderBolt or from any computer connected to the Internet.

Writing a non-Gmail message

I prefer to use Gmail on the phone. After all, the ThunderBolt is a Google phone. It tracks Gmail better than it tracks boring old email. Even so, you can compose a non-Gmail message on the ThunderBolt by using the Mail app. It works like this:

1. **Open the Mail app.**

2. **If necessary, choose a specific account.**

   Use the Select Account menu, as shown earlier, in Figure 10-1. If you don't select a specific account, mail is sent from the primary account.

3. **Touch the Compose Mail button.**

   See Figure 10-3 for its location.

4. **Craft the message.**

   Fill in the blanks just as you would when composing an email message on your computer.

5. **Touch the Send button to send the message.**

Copies of the messages you send in the Mail app are stored in the specific account's Sent folder. To see the folder, press the Menu soft button when viewing the account's inbox and choose Folders and then Sent.

- Mail is sent right away, unless no digital network is available. In that case, mail lingers in the Outbox folder until a signal is acquired.

- To remove a recipient from the To field, touch the button with their name or email address and choose the Remove Recipient command.

- To display the CC and BCC fields, press the Menu soft button when composing a message and choose the Show Cc/Bcc command.

- See the earlier section "Making a signature" to change the signature automatically attached to your outgoing messages.

- To cancel the message, touch either the Discard or Delete button.

- To save the message for later, touch the Save As Draft button.

- Drafts can be found in the Drafts folder: When viewing the inbox, press the Menu soft button and choose Folders, and then choose Drafts.

- Mail can get stuck in the outbox. A notification appears, which states that the account setting is incorrect. This type of error happens on Wi-Fi networks that may not use your same email (SMTP) server. The solution is to wait until another wireless connection is available, in which case the mail is sent automatically. (Until then, unsent messages wait in the outbox.)

Creating a new message for a contact

Perhaps the easiest way to compose a new email message is to find a contact and then create a message using that contact's email address. Quickly follow these steps:

1. **Open the People app.**

2. **Locate the contact to whom you want to send an electronic message.**

   Review Chapter 5 for ways to hunt down contacts in a long list.

3. **Choose an Email option from the contact's Action area.**

   The command to look for is Send Mail. When multiple Send Mail commands exist, you can check beneath each one to see which email address is specified.

4. **In the Complete Action Using dialog box, choose either Gmail or Mail to send a Gmail or Email message, respectively.**

   At this point, creating the message works as described in the preceding sections; refer to them for additional information.

Adding a message attachment

Any email message you create can sport an attachment, a photo or movie, or another type of file on your ThunderBolt. The operation to add an attachment to the message works similarly to the way it works on your computer. As a bonus, it's pretty much identical in both the Gmail and Mail apps:

1. **Press the Menu soft button and choose the Attach command.**

2. **Choose an app in which to fetch the attachment.**

   The Mail app features a much richer field of possibilities for attaching stuff.

3. **Find the media, or whatever, to attach to the message.**

   What happens at this point depends on the app, but basically you're using the app to hunt down something to attach.

4. **Continue composing the message after the file or media is attached.**

Another approach to sending an attachment is to start first by finding the attachment. Open the Gallery to find a picture or video, or open any other app in which you access or create media. The most common way to attach something is to long-press it or press the Menu soft button and look for the Share command. From the Share menu, you can choose Gmail or Mail to complete the message.

See Chapter 15 for more information on the Gallery.

The Portable World Wide Web

1 was blown away the first time I accessed the Internet on my ThunderBolt over a 4G connection. The information flew onto the screen with such laser-like velocity that I had to warn the other folks in the coffee shop not to walk through the signal.

Thanks to the ThunderBolt's über-fast 4G LTE data connection, you can use your phone to browse the web with blistering swiftness. It's the entire World Wide Web experience, but it happens in the palm of your hand wherever you go. Well, wherever your phone can find a data connection. The action also takes place on the phone's teensy screen. So the web experience on the ThunderBolt is speedy and tiny, like an Alka-Seltzer.

As long as you have 4G service, using the ThunderBolt's cellular data connection is just as fast, if not faster, than using the typical Wi-Fi connection. Even so:

↳ If possible, activate the ThunderBolt's Wi-Fi connection before you venture out on the web. Though you can use the phone's cellular data connection, Wi-Fi access incurs no data usage charges.

✔ Many places you visit on the web can instead be accessed directly and more effectively by using specific apps. Facebook, Gmail, Twitter, and YouTube, and potentially other popular web destinations, have apps that are either preinstalled on the phone or can be downloaded for free from the Android Market.

Web Browsing on Your Phone

The World Wide Web isn't as strange and unfamiliar as it was 20 years ago. What might be strange and familiar to you is using the web on a cell phone. Therefore, consider this section to be your mobile web orientation.

Getting on the web

Your ThunderBolt's web browsing app is named *Internet*. That name seems general, but HTC didn't bother calling me when it came time to name its phone's web browser app. To avoid confusion with the massive thing that is the actual Internet, I refer to the web browser on your phone as the *Internet app*.

The Internet app (See? — I'm doing it now) dwells as a shortcut icon, found on the Home screen. Figure 11-1 illustrates the Internet app's rather simple interface.

Here are some handy ThunderBolt web browsing tips:

✔ Pan the web page by dragging your finger across the touchscreen. You can pan up, down, left, or right.

✔ Double-tap the screen to zoom in or zoom out, though not every web page is zoomable.

✔ Pinch the screen to zoom out, or spread two fingers to zoom in.

✔ You can orient the phone horizontally to read a web page in Landscape mode. Then you can spread or double-tap the touchscreen to make teensy text more readable.

✔ The Internet app isn't the only way to surf the web on your phone. Another popular app is Dolphin Browser, which is available at the Android Market. See Chapter 18.

Looking at a web page

To visit a web page, type its address into the Address box (refer to Figure 11-1). You can also type a search word, if you don't know the exact address of a web page. Press the Enter key on the onscreen keyboard or the Go button by the Address box to search the web or visit a specific web page.

Figure 11-1: The Internet app.

If you don't see the Address box, swipe your finger so that you can see the top of the window, where the Address box lurks.

You "click" links on a page by touching them with your finger. If you have trouble stabbing the right link, zoom in on the page and try again.

- ✔ To reload a web page, touch the Refresh button on the right end of the Address bar.

- ✔ To stop a web page from loading, touch the X that appears to the right of the Address bar. The X replaces the Refresh button and appears only when a web page is loading.

Going forward and backward

menu Most of the action in the Internet app begins by pressing the Menu soft button. You'll find the basic web browsing commands located on the menu:

To return to a web page, press the Menu soft button and choose the Back command. Or, you can press the Back soft button.

To go forward, and return to the page you were visiting before you touched the Back button, press the Menu soft button and choose the Forward command.

Return to the Home page by pressing the Menu soft button: Choose the More command, and then choose Home.

To review the long-term history of your web browsing adventures, press the Menu soft button and choose More and then History. Your web browsing history appears, as shown in Figure 11-2.

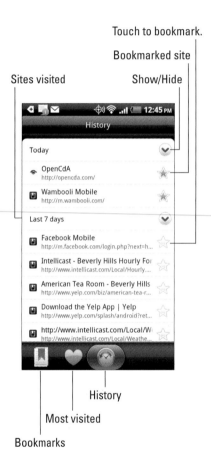

Touch to bookmark.

Bookmarked site

Sites visited   Show/Hide

History

Today

OpenCdA
http://opencda.com/

Wambooli Mobile
http://m.wambooli.com/

Last 7 days

Facebook Mobile
http://m.facebook.com/login.php?next=h...

Intellicast - Beverly Hills Hourly For
http://www.intellicast.com/Local/Hourly....

American Tea Room - Beverly Hills
http://www.yelp.com/biz/american-tea-r...

Download the Yelp App | Yelp
http://www.yelp.com/splash/android?ret...

http://www.intellicast.com/Local/W
http://www.intellicast.com/Local/Weathe...

History

Most visited

Bookmarks

Figure 11-2: The Internet app's history list.

To view a page you visited weeks or months ago, you can choose a web page from the History list. Touch the Show/Hide button, as illustrated in the figure, to see where you've been on previous days.

To clear the History list, press the Menu soft button and choose the Clear History command. It's the only command.

To remove an individual item from the history list, long-press the entry and choose the Remove from History command from the pop-up menu.

Working with bookmarks

Bookmarks are those electronic breadcrumbs you can drop as you wander the web. Need to revisit a website? Just look up its bookmark. This advice assumes, of course, that you bother to create (I prefer *drop*) a bookmark when you first visit the site. Here's how it works:

1. **Visit the web page you want to bookmark.**

2. **Press the Menu soft button.**

3. **Choose the Add Bookmark command.**

4. **If necessary, edit the bookmark name.**

   The bookmark is given the web page name, which might be kind of long. I usually edit the name to a shorter one, which is more visible on the Bookmarks screen, shown in Figure 11-3.

5. **Touch the Add button.**

After the bookmark is set, it appears in the list of bookmarks, as shown in Figure 11-3. To see this screen, press the Menu soft button and choose the Bookmarks command.

A helpful way to create bookmarks is to view the History screen and touch the Star button by a website name, as illustrated in Figure 11-2. You can also bookmark sites on the Most Visited screen: Touch the Heart button (refer to Figure 11-2) and long-press a website's thumbnail image. From the pop-up menu, choose Add Bookmark.

- ✔ To visit a bookmark, choose it from the Bookmarks list.

- ✔ Remove a bookmark by long-pressing its entry in the Bookmarks list. Choose the command Delete Bookmark. The bookmark is gone.

- ✔ Bookmarked websites can also be placed on the Home screen: Long-press the bookmark thumbnail and choose the command Add Shortcut to Home.

- ✔ The MyBookmarks app, obtained from the Android Market, can import your Internet Explorer, Firefox, and Chrome bookmarks from your Windows computer into the ThunderBolt. See Chapter 18 for more information on the Android Market.

Bookmark name

Web page preview

Swipe down for more bookmarks.

Figure 11-3: The Bookmarks screen.

Subscribing to a news feed

 A better option than setting a bookmark is to subscribe to a web page. Some pages, especially blogs and news sources, feature RSS feeds. When they do, you see the RSS subscription icon on the left side of the address bar, as shown in the margin.

To subscribe to the web page's news feed, touch the Subscription button. On the Add Subscription screen, place a green check mark by the news feeds you want to subscribe to. Touch the Add button to add the news feeds.

You read your web page subscriptions by opening the News app, which is found on the All Apps screen. Your RSS subscriptions are shown on the Subscriptions screen, with any new articles flagged by a number in a green circle. Touch the subscription feed to read more information or peruse the new articles, illustrated in Figure 11-4.

RSS stands for *Really Simple Syndication,* a method of sharing frequently updated web pages.

Open a new window.   Close window.

'Nuther window   Swipe window   'Nuther window
left or right.

Figure 11-4: Working with browser windows.

Using multiple web page windows

The Internet app sports more than one window, which is a feature you find on computer web browsers. You can open web pages in new windows and manage these multiple web page windows on the ThunderBolt. You have several ways to do it:

- ✏ *To open a new browser window,* press the Menu soft button and choose Windows to see the Windows screen. Touch the Plus button, illustrated in Figure 11-4. The new window opens, using the home page that's set for the Internet app.

- ✏ *To open a link in another window,* long-press the link. Choose the Open in New Window command from the menu that appears.

- ✏ *To open a bookmark in a new window,* long-press the bookmark and choose the command Open in New Window.

To access multiple windows, press the Menu soft button and choose the Windows command. When multiple windows are open, you see a screen similar to the one shown in Figure 11-4. Otherwise, you see only the lone window hanging there.

Switch between windows by choosing one from the Windows screen (refer to Figure 11-4). Swipe left and right to peruse the windows. Touch a window to display its full-screen content.

Close a window by touching the X button, illustrated in Figure 11-4.

See the section "Setting the home page," later in this chapter, for information on setting the Internet app's home page.

Finding stuff on the web

Google is synonymous with searching, so what better way to look for things on the web than go use the Google widget, often found floating on the Home screen, just to the left of the main Home screen, and shown in Figure 11-5. Use the Google widget to type something to search for, or touch the Microphone button to dictate what you want to find on the Internet.

Figure 11-5: The Google widget.

 To search for something anytime you're viewing a web page in the Internet app, press the Search soft button. Type the search term into the box. You can choose from a list of suggestions.

To find text on the web page you're looking at, as opposed to searching the entire Internet, follow these steps:

1. **Visit the web page where you want to find a specific tidbit o' text.**
2. **Press the Menu soft button.**
3. **Choose More and then Find on Page.**
4. **Type the text you're searching for.**
5. **Use the left or right arrow button next to the Search box to locate the text on the page — backward or forward, respectively.**

   The found text appears highlighted in green.
6. **Touch the Back soft button when you're done searching.**

See Chapter 22 for more information on widgets, such as the Google widget.

✏ The Google Search app, found in the All Apps screen, works like a full-screen version of the Google widget. Use the app to search the web.

✏ If you touch the G button on the Google widget, you see where the widget searches for information. Touch the Gear icon on that screen to refine the search.

✏ The Quick Lookup app can search Google, Wikipedia, or YouTube, translate text, or look up words in the Google online dictionary. This app can also be found on the All Apps screen.

Sharing a page

There it is! That web page you just *have* to talk about to everyone you know. The gauche way to share the page is to copy and paste it. Because you're reading this book, and even though you may not recognize what *gauche* means, you know the better way to share a web page. Heed these steps:

1. **Long-press the link or bookmark you want to share.**

2. **Choose the command Share Link.**

   A pop-up menu of places to share appears, looking similar to Figure 11-6. The variety and number of items on the Share Via menu depend on the applications installed on your phone.

Figure 11-6: Options for sharing a web page.

3. **Choose a method to share the link.**

   For example, choose Mail to send the link by mail, or Messages to share via a text message.

4. **Do whatever happens next.**

   Whatever happens next depends on how you're sharing the link: Compose the email or text message, for example. Refer to various chapters in this book for specific directions.

The best way to share a web page is to share a link to the web page. Ditto for YouTube videos and other media: Links are text. They take no time to send, consume only a tiny sip of data, and are more welcome by experienced Internet users.

To share the current page, the one you're viewing, press the Menu soft button and choose More and then Share Page.

Internet Downloading

Most people use the term *download* to refer to copying or transferring a file or other information. This term is technically inaccurate, but the description passes for social discussion.

Officially, a *download* is the transfer of information over a network from another source to your gizmo. For your ThunderBolt, the network is the Internet, and the other source is a web page.

- ✔ When the ThunderBolt is downloading information, you see the Downloading notification. Unlike the artwork shown in the margin, the arrow on the notification icon scrolls down repeatedly while the item is being downloaded. (I tried to get the Wiley Production Department to animate the margin art, but all I got were strange looks.)

- ✔ There's no need to download program files to the ThunderBolt. If you want new software, you can obtain it from the Android Market, covered in Chapter 18.

- ✔ The opposite of downloading is *uploading* — the process of sending information from your gizmo to another location on a network.

Stealing an image from a web page

The simplest thing to download is an image from a web page. It's cinchy: Long-press the image. You see a pop-up menu appear, from which you choose the command Save Image.

To view images you download from the web, you use the Gallery app. Downloaded images are saved in the All Downloads album.

    ✒ Refer to Chapter 15 for information on the Gallery.

    ✒ Technically, the image is stored on the phone's MicroSD card. The image can be found in the `download` folder.

Downloading a file

The web is full of links that don't open in a web browser window. For example, some links automatically download, such as links to PDF files or Microsoft Word documents or other types of files that can't be displayed by a web browser. To save this type of document on the ThunderBolt, long-press the link and choose the Save Link command from the menu that appears.

    ✒ You can view the saved file by referring to the Download History screen. See the next section.

    ✒ Some links may automatically download; clicking the link downloads the file instead of opening it in a new window.

    ✒ The Adobe Reader app is used on the ThunderBolt to display PDF files.

Reviewing your downloads

The Internet app keeps a list of all the stuff you download from the web. To review your download history, follow these steps:

1. **Press the Menu soft button.**

2. **Choose the More command and then Downloads.**

    The Download History list appears, similar to the one shown in Figure 11-7. You can choose an item from the list to view it.

3. **To exit the Download Manager, press the Back soft button.**

Downloaded notification   Show/Hide

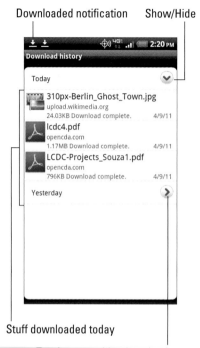

Stuff downloaded today

Touch to see older downloads.

Figure 11-7: The Download History list.

The Download History list keeps track of all the files you've downloaded using the Internet app, including web pages. To see older items, expand the list, as shown in Figure 11-7.

- To purge the list of downloaded files, press the Menu soft button while viewing the Download History list. Choose the Clear the List command. Clearing the list doesn't delete any files you've downloaded.

- To remove a single item from the Download History list, long-press it and choose the Delete command. Touch the OK button to confirm. Removing the item from the list also deletes the item you downloaded, so be careful.

- There are some things you can download that you cannot view. When it happens, you see an appropriately rude error message.

- You can quickly review any download by choosing the Download notification.

Internet App Configuration

More options and settings and controls exist for the Internet app than just about any other program I've used on the ThunderBolt. It's complex. Rather than bore you with every dang doodle detail, I thought I'd present just a few of the options worthy of your attention.

Setting the home page

The *home page* is the first page you see when you start the Internet app, and it's the first page that's loaded when you fire up a new window. To set your home page, heed these directions:

1. **Browse to the page you want to set as the home page.**
2. **Press the Menu soft button.**
3. **Choose More and then Settings.**

   A massive list of options and settings appears.

4. **Choose Set Home Page.**
5. **Choose the Use Current command.**

   The home page is set.

Remember: It's your *home* page. You don't have to use your cellular provider's home page or any other page that's preset for you.

If you want your home page to be blank (not set to any particular web page), follow Steps 1 through 4 in this section, but in Step 5 choose the Set Home Page command. Type **about:blank** in the text box — the word *about,* a colon, and then the word *blank,* with no period at the end and no spaces in the middle. Touch the OK button to set a blank home page.

I prefer a blank home page because it's the fastest web page to load. It's also the web page with the most accurate information.

Changing the web's appearance

You can do a few things to improve the way the web looks on your ThunderBolt. First and foremost, don't forget that you can orient the device horizontally to see a wide view on any web page.

From the Settings screen, you can also adjust the zoom setting used to display a web page. Heed these steps when using the Internet app:

1. **Press the Menu soft button.**

2. **Choose More and then Settings.**

3. **Choose Default Zoom.**

4. **Select Close from the menu.**

5. **Press the Back soft button to return to the web page screen.**

The Close setting might not be "big" enough, so you can spread your fingers to zoom in on any web page.

Reviewing privacy and security settings

With regard to security, my advice is always to be smart and think before doing anything questionable on the web. Use common sense. One of the most effective ways that the Bad Guys win is by using *human engineering* to try to trick you into doing something you normally wouldn't do, such as click a link to see a cute animation or a racy picture of a celebrity or politician. As long as you use your noggin, you should be safe.

As far as the ThunderBolt's browser settings go, most of the security options are already enabled for you, including the blocking of pop-up windows (which normally spew ads).

If web page cookies concern you, you can clear them from the Settings window. Follow Steps 1 and 2 in the preceding section and choose the option Clear All Cookie Data. Touch the OK button to confirm.

You can also choose the command Clear Form Data to remove all traces of text you've typed on a web page.

If you're concerned about your ThunderBolt being stolen and thieves gaining access to your web page passwords, remove the check mark by the item Remember Passwords. Ditto for the item Remember Form Data. While you're at it, choose the command Clear Passwords and touch the OK button to confirm.

The Clear History command is used to purge the list of places you've been. And the Clear Cache command removes any downloaded remnants of images and web page text from the phone's memory.

12

Your Digital Social Life

*F*orget the bars, forget the club, forget bridge with Brantley, Rowena, and Bob. If you want to be a social butterfly these days, you need to be on the Internet. The web is now the hub of social activity, due to popular social networking sites such as Facebook and Twitter. Armed with your ThunderBolt, there's no need to be left out of your gossip circles when you're out and about with your phone. It's a snap to easily share all your personal, private, and intimate moments with all of humanity.

Your Friend Stream

The locus for social networking activity on your ThunderBolt is the preinstalled Friend Stream app. This app is found on the All Apps screen, and the Friend Stream widget is located on the Home screen that's second to the right of the main Home screen.

- The Friend Stream app gives you access to updates and information from Facebook and Twitter and the online photo-sharing website Flickr.

- You can also use Friend Stream to share your social networking status with Facebook or Twitter or both at the same time.

- Other, specific apps deal with social networking in addition to Friend Stream. They're covered elsewhere in this chapter. The Friend Stream app, however, is better integrated with other apps (such as People) on your ThunderBolt.

- Though you can access Facebook, Twitter, and Flickr on the web by using the Internet app, I highly recommend that you use Friend Stream, or other specific apps, as described in this chapter.

Configuring the Friend Stream app

The first time you run the Friend Stream app, you're asked to log in to your social networking sites: Touch the Yes button to begin the process. You see the Accounts & Sync screen, which lists the social networking sites Facebook and Twitter and the image-sharing site Flickr. (Other sites may be listed in the future.)

As with other Internet activities, set up and configure your social networking on the Internet before you tackle the topic on your phone.

- Set up Facebook: www.facebook.com

- Set up Twitter: www.twitter.com

- Set up a Flickr account: www.flickr.com

Setting up all these accounts is optional. If you use only one, configure just one.

Obey these steps to complete the account configuration process:

1. **Choose the account you want to log in to, such as Facebook.**

2. **Type your username or email address.**

3. **Type your password for the account.**

4. **Touch the Sign In button.**

   After signing in, you return to the Accounts & Sync screen to sign in to additional accounts.

5. **Repeat Steps 1 through 4 to set up your other social networking accounts.**

When you're done setting up accounts, press the Home soft button and then run the Friend Stream app again to get started.

✔ You can add more accounts later, if you like: After starting the Friend Stream app, press the Menu soft button and choose Settings, and then choose Accounts & Sync.

✔ It's okay to authorize access to your Flickr account by using the HTC Media Updater. This permission is necessary in order to fully integrate Flickr, and your online photos, with the ThunderBolt.

Merging your social networking contacts

 After you stir your social networking sites into the Friend Stream app, the ThunderBolt instantly culls your list of friends and mixes them into the People app. You probably see the Matched Contacts Suggestion notification icon appear on the status bar, as shown in the margin.

I recommend that you merge your contacts. See Chapter 8, the section about matching identical contacts, for the specifics.

After merging, you can access social networking information from within the People app. You see links to a contact's Facebook and Twitter profiles as well as status updates directly on the contact's Details screen. Further, the People app gladly steals your Facebook and Twitter contact photos for use as photos for your address book.

You can touch the Updates and Events button at the bottom of the contact's information screen to see current Facebook updates and Twitter tweets. Refer to Figure 8-2, in Chapter 8, for the location of the Updates and Events button.

This kind of integration between the Friend Stream and People apps is one reason why I recommend using Friend Stream either instead of or in addition to specific social networking apps.

Viewing the streams

After getting your accounts all squared away, you can use the Friend Stream app to peruse online updates from your social networking buddies. The main All Updates screen in the Friend Stream app is shown in Figure 12-1.

Twitter update

Facebook update

Type your status update.

All Updates

Status Updates

Photos and Videos

Links

Lists

Notifications

Status updates

Figure 12-1: Status updates for your friends.

The All Updates screen, shown in Figure 12-1, lists any kind of update that's sent to all social networking services to which you subscribe. Touching other buttons on this tab (refer to Figure 12-1) refines the types of updates you see.

- ✔ Scroll the list to see what everyone is up to.

- ✔ Wee li'l icons in the pictures (as shown in Figure 12-1) reveal the social networking source: Facebook, Flickr, or Twitter.

- ✔ Touch an item to see more information or details or to visit a link.

- ✔ If you touch a *tweet* (a Twitter update), you're taken to the Peep app, which displays all your Twitter feeds. See the later section "Peep Your Thoughts on Twitter" for information on the Peep app.

- ✔ If you wander too far, press the Back soft button to return to the Friend Stream app.

- ✔ To see the latest Friend Stream updates, press the Menu soft button and choose the Refresh command.

- ✔ As it's normally configured, the Friend Stream app automatically refreshes itself every time it's started.

✔ The same update information found in the Friend Stream app is echoed on the Friend Stream widget, found on the second Home screen to the right of the main Home screen. To get to the Friend Stream app from the Friend Stream widget, touch the arrow button, as shown in the margin.

Adding a status update

There's no point in joining a social network if you're not going to be social. Using the Friend Stream app, sharing the most intimate details of your life with the entire online universe is as simple as it can be potentially embarrassing. Follow these steps:

1. **Touch the What's On Your Mind text box.**

   Refer to Figure 12-1 for its location.

   After touching the box, you see the What's on Your Mind screen, illustrated in Figure 12-2.

Figure 12-2: Sharing an update on Friend Stream.

2. **Type something pithy, profound, perfunctory, personal, perceptive, practical, poetic, political, private, perplexing, or positive — or just jot down a pun.**

   Lamentably, you're limited to only 140 characters.

3. **Touch the Update button to, optionally, choose from which social networking site to send the update.**

   If you don't choose a specific site, your status update is posted to all social networking accounts you've specified for use with the Friend Stream app.

4. **Touch the Attach button to, optionally, add an attachment to the update message.**

   You can attach a picture, a location, or an app recommendation (a link). Or, you can elect not to attach anything.

   See the next section for information on uploading a photo.

5. **Touch the Share button to spread your social networking message far and wide.**

After sharing the message, you return to the Friend Stream app, where you see your updates posted in just a few quick Internet seconds — just like all your buddies and followers.

  ✔ Another way to share information is to copy a link from the Internet app and then paste it into your status. See Chapter 11 for information on the Internet app, and Chapter 4 for the whole copy-paste ordeal.

  ✔ To best share web pages, look for the Facebook or Twitter Share button on the page. Touching the button offers an easy, simple way to share the page.

  ✔ Various Share buttons and menu items appear in many apps on your ThunderBolt. To share what you're looking at on the phone, choose the proper command from the Share menu: Facebook for HTC Sense to share on Facebook; Peep to share on Twitter; or Flickr for image sharing on Flickr.

  ✔ People on Twitter traditionally use URLs or link shortcuts instead of pasting in full web page links or images. To set the URL shortening server for the Friend Stream app, press the Menu soft button and choose URL Shortening Server. The Friend Stream app is preconfigured to use `bit.ly` as its URL shortening server.

  ✔ You can get around the 140-character limit when writing a Facebook status update, but you need to use the Facebook app instead of Friend Stream. See the section "The Facebook App," a little later in this chapter.

Uploading a photo

Let me be honest: The best way to upload a photo to Facebook, Flickr, or Twitter is to use the Gallery app, which is where the ThunderBolt keeps all the pictures you take. This topic is covered in Chapter 15.

- ✔ The Facebook app, covered in the next section, offers better integration of the ThunderBolt's camera with Facebook.

- ✔ For Twitter, the Peep app that comes with your ThunderBolt also features a handy Camera button. See the later section "Posting a picture to Twitter."

- ✔ The Camera button is hidden deep within the bowels of the Friend Stream app. I'd describe how to find it, but even after you find it, you have no simple mechanism for getting the picture back into Facebook or Twitter. Just see Chapter 15.

- ✔ Also see Chapter 14 for general information on using your phone as a camera.

The Facebook App

A more versatile way to access Facebook on your ThunderBolt is to obtain and use the Facebook app. It offers more specific and less limiting features than Friend Stream, covered earlier in this chapter.

As I mention earlier in this chapter, you should set up and configure your Facebook account using a computer on the Internet. Then you can access your Facebook account from the ThunderBolt without having to toil through the setup process.

Installing the Facebook app

The ThunderBolt doesn't come with a Facebook app, but you can get the official Facebook app for free from the Android Market. The app is your red carpet to Facebook's social networking kingdom.

To get the Facebook app, scan the QR code shown in the margin. Or, you can go to the Android Market and search for the Facebook for Android app. See Chapter 18.

The Facebook app is installed on your ThunderBolt when you see the Successful Install notification, shown in the margin. You can choose this notification to start the Facebook app or just refer to the next section.

Running the Facebook app

The first time you behold the Facebook app, you'll probably be asked to log in. Do so: Type the email address you used to sign up for Facebook and then type your Facebook password. Touch the Login button.

Proceed to sync your contacts, if asked; select the Sync All option, which brings in the names of all your Facebook friends to the phone's address book. Touch the Next button, and then touch Finish to begin using Facebook.

Eventually, you see the Facebook news feed or status update feed. To go to the main Facebook page, shown in Figure 12-3, press the Back soft button.

The Facebook app updates itself whenever it's loaded or every hour. To update the news feed immediately, press the Menu soft button while viewing the news feed and choose the Refresh command.

 When you're done using Facebook, press the Home soft button to return to the Home screen.

Update Status

Pending notifications

Recent photos and media

Figure 12-3: Facebook on the ThunderBolt.

The Facebook app continues to run, until you either sign out or turn off the ThunderBolt. To sign out of Facebook, press the Menu soft button and choose the Logout command.

- Refer to Chapter 22 for information on placing the Facebook app shortcut on the Home screen.

- Facebook comes with a Facebook widget that displays recent status updates and allows you to share your thoughts directly from the Home screen. See Chapter 22 for information on adding widgets to the Home screen.

- To see more details on an item in the news feed, choose the item. It appears on another screen, along with any comments and more details.

- Whenever something new happens on Facebook, you see the Facebook notification appear, similar to the one shown in the margin. (Different icons are used for different Facebook activities.) Pull down the notifications and choose the Facebook item to see what's up.

- Notifications also appear at the bottom of the main Facebook page, as shown in Figure 12-3.

Updating your status

The primary thing you live for on Facebook, besides having more friends than anyone else, is to update your status. It's the best way to share your thoughts with the universe, far cheaper than skywriting and far quicker than a message in a bottle.

To set your status on the ThunderBolt, follow these steps in the Facebook app:

1. Touch the Update Status button.

The button is shown in the margin but also appears atop the main Facebook screen, shown in Figure 12-3.

2. Type your comment into the What's On Your Mind text box.

You can also dictate the comment if you touch the keyboard's Microphone button.

3. Touch the Share button.

You can also set your status by using the Facebook widget on the Home screen, if the widget has been installed: Touch the What's On Your Mind text box, type your important news tidbit, and then touch the Share button.

Sending a picture to Facebook

One of the many things your ThunderBolt can do is take pictures. Combine this feature with the Facebook app and you have an all-in-one gizmo designed for sharing the various intimate and private moments of your life with the ogling throngs of the Internet.

 The key to sharing a picture on Facebook is to locate the wee Camera icon, which is found to the left of the What's On Your Mind text box on the Facebook app's News Feed page. Here's how to work the button and upload an image or a video to Facebook:

1. **Touch the Camera icon on the News Feed screen.**

   There's also a Camera button on the My Albums page, which you get to by touching the Photos icon on the main Facebook app screen (see Figure 12-3).

2. **Choose an option from the Upload Photo menu.**

   You have two options for uploading a picture:

   Choose from Gallery: If you choose this option, browse the Gallery to look for an existing picture that you want to upload. (See Chapter 15 for more information on how the Gallery app works.)

   Capture a Photo: Use the ThunderBolt's camera to snap a picture of whatever is around you. Touch the shutter button to snap the picture; touch Done to continue or the other button (with the Camera icon on it) to start over and take a new picture. (See Chapter 14 for more information on how to use the Camera app.)

   After selecting or taking a picture, you see the Upload Photo screen, shown in Figure 12-4.

3. **Optionally, choose an album by touching the Album button.**

    Unless you choose otherwise, the Facebook app uploads your picture to the Mobile album.

4. **Optionally, type a caption.**

   Touch the Add a Caption Here text box and the onscreen keyboard appears. Type or dictate a caption for your image. The caption is also uploaded to Facebook.

5. **Touch the Upload button.**

   The image is posted as soon as it's transferred over the Internet and digested by Facebook.

Send the picture to Facebook.

Type an overly clever caption.

Photo

Figure 12-4: Uploading an image to Facebook.

The image can be found as part of your status update feed or news feed, but it's also saved to whichever album you specified (refer to Step 2).

- ✔ You can use the Facebook app to view the image in Facebook, or you can use Facebook on any computer connected to the Internet.

- ✔ You can also use the Friend Stream app, which is sorely lacking when it comes to uploading pictures.

- ✔ Even though you can use the Camera app to shoot video, this feature is disabled when you go to take a picture for Facebook.

- ✔ Facebook also appears on the various Share and Share Via menus you find on the ThunderBolt. Choose the Facebook item to send to Facebook a copy of whatever it is you're looking at.

- ✔ Other chapters in this book give you more information about the various Share and Share Via menus and where they appear.

- ✔ The Facebook app uses the Facebook command on the Share and Share Via menus; Friend Stream uses the Facebook for HTC Sense command.

Peep Your Thoughts on Twitter

Twitter is a social networking site, similar to Facebook but far more brief and a bit more newsy. On Twitter, you write short spurts of text, or *tweets,* which express your thoughts or observations. You can also follow the tweets of others.

The ThunderBolt ships with the Twitter client known as the Peep app. The app is already tied into Friend Stream, so if you've configured Friend Stream for use with your Twitter account, you're pretty much set up and ready to go.

- ✔ As with other social networking sites, I recommend that you set up and configure your Twitter account using a computer connected to the Internet.

- ✔ Though you can access Twitter on the Internet using the Internet app on your ThunderBolt, I recommend using a specific Twitter app, such as Peep.

- ✔ You can use other Twitter apps in addition to Peep. They include the official Twitter app as well as the popular Twidroyd app. Both apps, and more Twitter clients and widgets, can be found at the Android Market. See Chapter 18.

Using Peep

Though you can read and share tweets using the Friend Stream app, the Peep app offers you an exclusive Twitter venue. It has none of that Facebook detritus to sweep over your screen like a flock of Dorito crumbs.

Start the Peep app by choosing it from the All Apps screen. If you haven't yet configured Twitter for your ThunderBolt, you may be asked to sign in: Do so.

The main Peep screen is shown in Figure 12-5. You use this screen to follow the latest news, share your thoughts, or engage in other activities, as illustrated in the figure.

The main tweet feed is shown in Figure 12-5. Touch the buttons at the bottom of the screen to see your mentions or direct messages or tweets marked as favorites.

You can stay logged in to Twitter as long as you're using your phone. You can quit the Twitter app by pressing the Home soft button at any time. Quitting Twitter doesn't sign you out, so you continue to receive tweets while you're doing other things with your phone.

Tweets

Touch to tweet.

All Tweets

Mentions

Direct Messages

Favorites

Figure 12-5: The Peep app.

To sign out of Twitter, follow these steps:

1. **Go to the All Tweets screen in the Peep app.**

   Refer to Figure 12-5.

2. **Press the Menu soft button.**

3. **Choose More and then Settings.**

4. **Choose Account Settings.**

5. **Touch the Sign Out button.**

6. **Touch the Yes button in the warning dialog box.**

New tweets and Twitter notifications aren't received by the Peep app when you're logged out.

Tweeting your thoughts

You have only 140 characters to compose a tweet on Twitter. Be brief. Be pithy. Be interesting and people will follow you, which is the goal of the whole thing.

To tweet, touch the What's Happening text box, shown in Figure 12-5. You see the What's Happening screen appear, as shown in Figure 12-6, which is where the tweet action takes place. Refer to the figure for exciting things you can tweet.

Tweet

Your brilliant tweet

Character count

Attach a location.

Attach your photo.

Figure 12-6: Creating a new tweet.

Touch the Update button to share your thoughts with both of your Twitter followers.

Also see the next section, on uploading pictures to Twitter.

If you press the Menu key while viewing the What's Happening screen, you see two nifty menu items:

Quick Text: Choose this item to see a list of common tweets. Touch a tweet to make that text appear in the Tweet text box.

Insert Smiley: Choose this item to see a palette of smiley icons, also known as *emoticons.* Choose one to insert it into your tweet.

You have only 140 characters for creating a tweet. The count includes spaces.

Posting a picture to Twitter

One option for creating a new tweet (refer to Figure 12-6) is to post a picture to Twitter. You can choose to either summon a picture from the Gallery or take a picture with the phone right then and there.

Twitter itself doesn't display pictures, other than your account picture. When you send a picture to Twitter, you use an image hosting service and then share the link, or URL, to the image. All this complexity is handled by the Peep app; you just follow the onscreen instructions. It works something like this:

1. **Press the Post Picture button.**

   Refer to Figure 12-6 for the button's location on the What's Happening screen.

2. **Choose From Camera to take a picture, or From Gallery to mull the photos already stored on your phone.**

   See Chapter 14 for directions on how to use the ThunderBolt camera.

   See Chapter 15 for specifics on how the Gallery works.

   After capturing or selecting the image, you see a URL or web page link to the image appear as your tweet.

3. **Touch the Update button to share the image's link with your Twitter followers.**

The Peep app appears on various Share and Share Via menus available in other apps on your ThunderBolt. You use these Share menus to send to Twitter a copy of whatever you're looking at.

Following others on Twitter

Only a tiny percentage of the folks who use Twitter bother to tweet. What's far more popular is following the tweets of others. From news tweets to the random thoughts of your favorite celebrities or politicians whom you despise, following people on Twitter using your ThunderBolt is a cinch.

The easiest way to follow someone is to find one of those Follow Me on Twitter links on a web page. Touch the link while browsing the web on your phone, which takes you to the Twitter mobile web page (not the Peep app). Touch the Follow button, and log in to Twitter on the web, if prompted.

Another way to follow someone on Twitter is to use the Peep app to search for the person or organization. Follow these steps:

1. **While using the Peep app, press the Search soft button.**

2. **Choose the search type from the Searchable Items menu.**

 3. **Refine your search.**

   Choose All to search tweets and topics; choose Tweets to search the text of tweets; or choose Users to look for specific people.

4. **Type the person's name, an organization name, or a topic.**

5. **Peruse the results to find whom or what you want to follow.**

 Or, if you see no specific results, touch the Search button, shown in the margin.

When you find a person you want to follow, press the Menu soft button and choose the Follow command. If you don't see that command, touch one of the topics or people displayed on the screen.

To unfollow someone, long-press one of their tweets and choose Show Contact. When the contact's information is visible, press the Menu soft button and choose the Unfollow command. No warning is displayed when you choose to unfollow someone.

Other Social Networking Opportunities

Your social networking options aren't limited to Facebook and Twitter, though they're the most popular. Seeing how the web goes through these fads, you can find a myriad of other social networking sites. It seems like a new one pops up every week.

Beyond Facebook and Twitter, other social networking sites include, but aren't limited to

- ✔ Google Buzz
- ✔ Google Latitude
- ✔ LinkedIn
- ✔ Meebo
- ✔ MySpace

I recommend first setting up the social networking account on your computer, similar to the way I describe earlier in this chapter for Facebook and Twitter. After that, obtain an app for the social networking site by using the Android Market. Set up and configure the app on your ThunderBolt to connect with your existing account.

- ✔ Social networking sites may have special Android apps you can install on your ThunderBolt, such as the MySpace Mobile app for MySpace.

- ✔ Google Latitude is now part of the Maps app. See Chapter 13 for more information on using the Maps app on your phone.

- ✔ As with Facebook and Twitter, you may find your social networking apps appearing on various Share menus on the phone. That way, you can easily share your pictures and other types of media with your online social networking pals.

Part IV
Superphone Duties

The 5th Wave   By Rich Tennant

"You can do a lot with a ThunderBolt, and I guess dressing one up in G.I. Joe clothes and calling it your little desk commander is okay, too."

In this part . . .

The ancient Greek gods were once well integrated into the culture. Now they're relegated to high school humanities classes, perhaps a category on *Jeopardy!*, and the occasional special-effects-laden Hollywood fantasy film. Replacing the old gods are the new gods, though we don't call them gods. We call them *superheroes*.

Superheroes such as Superman, Batman, Spiderman, and others have replaced the Greek gods as the basis of our culture's mythos. Like gods, superheroes have special abilities and powers, or just a handy set of tools that provides them with a nearly infinite ability to fight the bad guys and entertain. In that way, your ThunderBolt phone is similar to a superhero, primarily because it can do so much more than a mere mortal phone.

There's a Map for That

....

In This Chapter

▶ Using the Maps app

▶ Displaying special map options

▶ Searching for people, places, and things

▶ Getting to your destination

▶ Finding a business or contact

▶ Navigating with the ThunderBolt

▶ Adding a Home screen navigation shortcut

....

*T*he group was stunned and disoriented. The world rocked, as gravity tugged left and right instead of down.

Eventually, Chaz righted himself; his nausea abated. "How is Dr. Cornelius going to sell teleportation to the masses with that kind of aftereffect?" he grunted.

Deb agreed. "Not only that — where are we? It would be nice if teleportation had a definite destination."

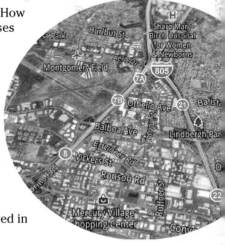

"It doesn't matter," replied Dan. "I have my trusty ThunderBolt phone. Using the Maps app, I can tell you exactly where we are. I can even quickly locate a decent Asian restaurant. Who's up for kimchi?"

Thanks to the ThunderBolt phone, its GPS abilities, plus the handy Maps app, you too can quickly locate yourself, friends, places to go, and more. It's all covered in this chapter.

✔ The Maps app has two sister apps, both of which rely upon the Maps app to perform specific duties: The Navigation app handles the chore of getting from here to there, similar to the way a GPS car-navigation system works; the Places app helps you search for specific things when you're out and about.

✔ Both the Navigation and Places apps can be accessed from within the Maps app and from the All Apps screen.

✔ Another app that uses your location is Footprints, though it's more of a travelogue or share-your-journey type of app and not about finding your way.

✔ Google frequently updates the Maps app. For information on recent updates, please refer to my website:

```
www.wambooli.com/help/phone
```

You Are Here

The Maps app combines the ThunderBolt's ability to read global positioning system (GPS) satellites with a vast database of streets and satellite images. It's like a marriage between the traditional street map and the yellow pages, though there's nothing to improperly fold or to attempt to tear in half.

✔ The ThunderBolt's GPS is the same technology used by car navigation toys as well as by handheld GPS gizmos.

✔ When the ThunderBolt is using the GPS radio, you see the GPS Is On status icon appear.

✔ The Maps app works better when you turn on the wireless networking, or *Wi-Fi*. The app bugs you when you run it and Wi-Fi is disabled.

✔ The ThunderBolt lets you know when various applications access Location features. The warning is nothing serious: The Android operating system is just letting you know that software accesses the device's physical location. Some folks may view this feature as an invasion of privacy; hence the warnings. I see no issue with letting the phone know where you are, but I understand that not everyone feels that way. If you'd rather not share location information, simply decline access when you're prompted.

Unfolding the Maps app

You start the Maps app by choosing its icon from the All Apps screen. If you're starting the app for the first time, or if it's just been updated, you can read the What's New screen; touch the OK button to continue.

The ThunderBolt uses its GPS abilities to zero in on your current location on Planet Earth. You see the location as a blinking blue triangle, as shown in Figure 13-1. If your location isn't exact, the Maps app displays a blue circle around the triangle, which gives your approximate location.

Where you are on the map

Compass pointer   Places

Find stuff.   Layers

GPS radio is on.   Location

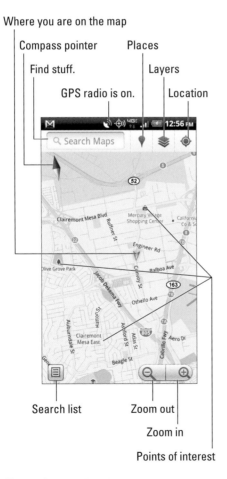

Search list   Zoom out

Zoom in

Points of interest

Figure 13-1: Your location on a map.

If you don't see your location on the map, touch the Location button (refer to Figure 13-1). Or, you can touch the Location button at any time to return to your current location from elsewhere on the map.

Here are some fun things you can do when viewing the map:

Zoom in: To make the map larger (to move it closer), touch the Zoom In button, double-tap the screen, or spread your fingers on the touchscreen.

Zoom out: To make the map smaller (to see more), touch the Zoom Out button or pinch your fingers on the touchscreen.

Pan and scroll: To see what's to the left or right or at the top or bottom of the map, drag your finger on the touchscreen; the map scrolls in the direction you drag your finger.

Rotate: Using two fingers, rotate the map clockwise or counterclockwise. Touch the Compass Pointer, shown in Figure 13-1, to reorient the map with north at the top of the screen.

Perspective: Tap the Location button to switch to Perspective view, where the map is shown at an angle. Touch the Perspective button again to return to a flat map view or, if that doesn't work, touch the compass pointer.

The closer you zoom in to the map, the more detail you see, such as street names, address block numbers, and businesses and other sites. And those tiny little specs you see on the screen? Yeah, that's dust, not people. See Chapter 23 for information on cleaning the screen.

> ✔ The blue triangle (refer to Figure 13-1) shows in which general direction the phone is pointing.
>
> ✔ Perspective view can be entered for only your current location.
>
> ✔ When all you want is a virtual compass, similar to the one you lost as a kid, you can get the Compass app from the Android Market. See Chapter 18 for more information on the Android Market.

Adding layers

You add or remove information from the Maps app by applying or removing layers. A *layer* can add detail, information, or other fun features to the basic street map, as shown in Figure 13-2.

The key to accessing layers is to touch the Layers button, illustrated in Figure 13-2. Choose an option from the Layers menu to add that information to the Map's app display.

Points of interest

Touch to exit Perspective view.

Your location   Layers

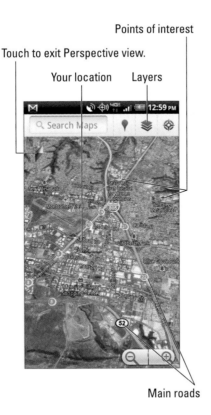

Main roads

Figure 13-2: The Satellite layer.

For example, in Figure 13-2 you see the Satellite layer displayed. You can still see the Street layer, but more detail is available, thanks to the satellite imagery.

You can add another layer by choosing it from the Layers menu, but keep in mind that some layers obscure others. For example, the Terrain layer overlays the Satellite layer so that you see only the Terrain layer.

To remove a layer, choose it from the Layers menu; any active layer appears with a green check mark to its right.

TIP

 ✓ To see 3D terrain, touch the Layers button and choose Terrain. You can combine Terrain view with the Perspective trick (described in the preceding section) to see the lay of the land.

 ✓ The Traffic layer, available from the Layers menu, shows current traffic conditions color-coded on the street map. Traffic information may not show up for every location, however.

 ✓ Also see the later section "Finding out where you are" for details about the Street View feature.

Spicing things up with Labs

Additional options for the Maps app are available from Google Labs. In fact, most options now available to the Maps app had their birth in Labs.

To see what's cooking in Labs for the Maps app, press the Menu soft button, choose More, and then choose Labs. You see a list of various features you can add to the Maps app, along with a description of what they do. Choose an item to add it to the app.

The variety of items available in Labs depends on what's being developed by Google. The list may change, and new items may be added or items removed or other items promoted to regular features in the Maps app. (That's why the Maps app is frequently updated.)

Search Your World

The true power of the Maps app lies in its powerful ability to search locations. You can use the app to locate where you are, see where things are nearby, or search for people or businesses, or you may have the Maps app tell you how to get there. It's one of my favorite apps on the ThunderBolt. This section explains how all this stuff works without getting you lost.

Finding out where you are

Your location on the Maps app screen is shown as a blue triangle (refer to Figure 13-1). If you don't see the triangle on the screen, touch the Location button. It helps you find where you are. But, seriously: *Where are you?* If you want to contact a friend to pick you up, you can't just explain that you're a blue triangle on the screen. The friend won't believe you, no matter how much you promise that you haven't been drinking.

To find out where you are, at your current street address, long-press your location on the Maps app. Or, you can long-press any location to find out the location's specific address. Up pops a bubble, similar to the one shown in Figure 13-3, that gives your approximate address.

If you touch the address bubble (refer to Figure 13-3), you see a screen full of interesting things you can do, as shown in Figure 13-4.

Touch the bubble to see more info.

Long-press a location to see the address.

Figure 13-3: Finding an address.

The What's Nearby command displays a list of nearby businesses or points of interest. Some of them show up on the screen, and others are available by touching the What's Nearby command.

Choose the Search Nearby item to use the Search command to locate businesses, people, or points of interest near the given location.

The Report a Problem command doesn't connect you with the police; instead, it's used to send information back to Google regarding an improper address or another map malfunction.

What's *really* fun to play with is the Street View command. Choosing this option displays the location from a 360-degree perspective. In Street view, you can browse a locale, pan and tilt, or zoom in on details to familiarize yourself with an area, for example — whether you're familiarizing yourself with a location or planning a burglary.

Your location   Phone a business or location.

Return to the map.   Mark the location as a favorite.

Get directions.   Street view preview

Street view

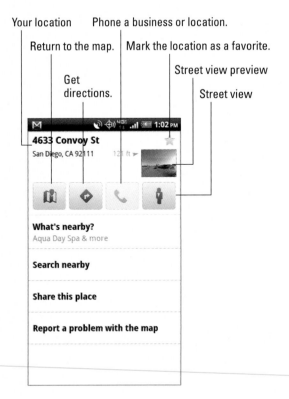

Figure 13-4: Things to do with a location.

Press the Back button to return to regular Map view from Street view.

Finding people and places

The Maps app can work like a powerful web search engine. Unlike when you're finding things on the Internet, however, you use the Maps app to locate things in the real world.

The secret to searching for a location is to press the Search soft button. Or, you can simply touch the Search Maps text field atop the Maps app window. The following sections describe how to search for various items.

Look for a specific address

To locate an address, type it in the Search Maps text box; for example:

```
1 Legoland Drive, Carlsbad, CA 92008
```

Touch the Search button to the right of the Search Maps text box and that location is then shown on the map. The next step is getting directions, which you can read about in the later section "Getting directions."

✔ You don't need to type the entire address. Oftentimes, all you need is the street number and street name and then either the city name or zip code.

✔ You can use dictation by touching the Microphone button and then speaking the address.

✔ If you omit the city name or zip code, the ThunderBolt looks for a matching address near your current location.

Find a type of business, restaurant, or point of interest

You may not know an address, but you know when you desperately need some green tea. Maybe you need a hotel or gas station. To find a business entity or a point of interest, type its name in the Search box; for example:

```
Green tea
```

This command flags coffee shops or a *palais du thé* near your current location. Or, you can be specific and look for locations elsewhere by specifying the city name, district, or zip code, such as

```
tea 92123
```

After typing this command and touching the Search button, you see a smattering of tea restaurants (or coffeehouses) found in my old neighborhood in San Diego, as shown in Figure 13-5.

To see more information about a result, touch its cartoon bubble, such as the one for Broadway Coffee, shown in Figure 13-5. The screen that appears offers details, plus perhaps even a website address and phone number. You can press the Get Directions button (refer to Figure 13-4) to get driving directions; see the later section "Getting directions."

✔ Each letter or dot on the screen represents a search result.

✔ Use the Zoom controls or spread your fingers to zoom in to the map.

menu

✔ You can create a contact for the location, keeping it as a part of your Contacts list: After touching the location balloon, touch the Menu button (shown in the margin). Choose the command Add As a Contact. The contact is created using data known about the business, including its location and phone number and even a web page address — if the information is available.

✔ When the location's information page shows a Phone button, you can touch that button to phone the place. Make a reservation. Ask whether they have loose-leaf sencha.

✔ Touching the Search List button on the Maps screen displays all search results in list format. Refer to Figure 13-1 for the location of the Search List button.

Search nearby

Maybe you don't know what you're looking for. Maybe you're like my teenage sons, who stand in front of the open refrigerator, waiting for the sandwich fairy to hand them a snack. The Maps app features a sort of I-don't-know-what-I-want-but-I-want-something fairy. It's the Places command.

Figure 13-5: Search results for *tea* in Kearny Mesa.

Touch the Places button (refer to Figure 13-4) to see a list of places near you: restaurants, coffee shops, bars, hotels, attractions, and more, as shown in Figure 13-6. Touch an item to see matching locations in your vicinity.

Go to or return to Maps app.

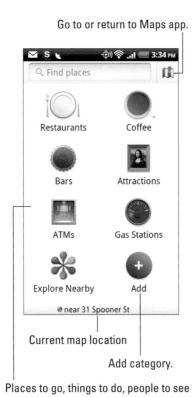

Current map location

Add category.

Places to go, things to do, people to see

Figure 13-6: Places to go, things to see.

You can also use the Places app to directly visit the Places screen (refer to Figure 13-6). You find the Places app on the All Apps screen.

> ✔ Use the Add button to create another search category, such as Candy Store, Speak-easy, or Industrial Espionage Retail Outlet.

> ✔ To remove a category, long-press its icon and choose the Remove command.

Find a contact

You can hone in on where your contacts are located by using the map. This trick works when you've specified an address for the contact — either home or work or another location. If so, your ThunderBolt can easily help you find that location or even give you directions. Heed these directions:

1. **Display information about a contact.**

   Use the People app, as described in Chapter 8.

2. **Touch the contact's address.**

   The address has text above it that reads *View Home Address* or *View Work Address.*

3. **If you see the Complete Action Using menu, choose Maps.**

   The contact's location appears in the Maps app.

You can touch the address bubble to get directions or read more information about the location. Refer to sections elsewhere in this chapter for the specifics.

- When your ThunderBolt features more than one map search application, you see the Complete Action Using menu (refer to Step 3).

- Not every contact has address information available, though there's nothing to stop you from editing your contacts and adding their address information.

- See Chapter 8 for more information on using the People app.

Getting directions

One command associated with locations on the map is Get Directions. I suppose it's the opposite of the Get Lost command. Here's how to use it:

1. **Touch a location's cartoon bubble displayed by an address, a contact, or a business or from the result of a map search.**

2. **Touch the Get Directions button.**

   See the next section for information on the Navigation options.

3. **Choose Get Directions.**

   You see the directions listed, as shown in Figure 13-7. The Maps app has already chosen your current location (shown as My Location in the figure) as the starting point and the location you searched for or are viewing on the map as the destination.

Choose a new starting
location or destination.

Distance to destination

Switch to Navigation.

Approximate
time to
destination

Transportation method   Map

Directions   Always good to know

Figure 13-7: Going from here to there.

You can follow the directions on the screen or touch the Map button (refer to
Figure 13-7) to see the path you need to take, as illustrated on the map. Zoom
in or out to see more or less detail.

- To receive vocal directions, touch the Navigation button or just read the
  next section.

- Touch the From or To field to change the point of origin or the destination, though I admit that it's easier to simply start over.

- The Transportation Method button (refer to Figure 13-7) displays a
  menu from which you can choose a different way to travel, such as on
  foot, by bicycle, or via public transportation (transit).

✓ In Figure 13-7, the Maps app notes toll roads on the specified route. As you travel, you can choose alternative, non-toll routes, if they're available. You're prompted to switch routes during navigation; see the next section.

✓ The Maps app may not give you the perfect directions, but for places you've never been, it's a useful tool.

Navigating to your destination

Lists are *so* 20th century. I don't know why anyone would bother, especially when the ThunderBolt features a digital copilot, in the form of voice navigation.

To use Navigation, choose the Navigation option from any list of directions. Or, touch the Navigation button, as shown in Figure 13-6. You can also enter the Navigation app directly by choosing it from the All Apps screen, though then you must input (or speak) your destination, so it's easier to start in the Maps app.

In Navigation mode, the ThunderBolt displays an interactive map that shows your current location and turn-by-turn directions for reaching your destination. A digital voice tells you how far to go and when to turn, for example, and gives you other nagging advice — just like a backseat driver, albeit an accurate one.

After choosing Navigation, sit back and have the ThunderBolt dictate your directions. You can simply listen or just glance at the phone for an update on where you're heading.

menu To stop Navigation, press the Menu soft button and choose the Exit Navigation command.

✓ To remove the navigation route from the screen, exit Navigation and return to the Maps app. Touch the Layers button to bring up the Layers menu. Touch the Clear Map button.

✓ When you tire of hearing the Navigation voice, press the Menu soft button and choose the Mute command.

✓ I refer to the navigation voice as *Gertrude*.

✓ You can press the Menu soft button while navigating and choose Route Info to see an overview of your journey.

▶ When viewing the Route Info screen, touch the Gears button to see a handy pop-up menu. From this menu, you can choose options to modify the route to stay off major highways or avoid toll roads.

▶ The neat thing about Navigation is that whenever you screw up, a new course is immediately calculated.

▶ In Navigation mode, the ThunderBolt consumes a lot of battery power. I highly recommend that you plug the phone into your car's power adapter ("cigarette lighter") for the duration of the trip.

Adding a navigation shortcut to the Home screen

When you visit certain places often — such as the liquor store — you can save yourself the time you would spend repeatedly inputting navigation information, by creating a Navigation shortcut on the Home screen. Here's how:

1. **Long-press a blank part of the Home screen.**

   The Personalize menu appears.

2. **Choose Shortcut.**

3. **Choose Directions & Navigation.**

4. **Type a contact name, address, destination, or business name in the text box.**

   As you type, suggestions appear in a list. You can choose a suggestion to save yourself some typing.

5. **Choose a traveling method.**

   Your options are car, public transportation, bicycle, and foot (even though the icon of the "on foot" guy seems to have no feet).

6. **Scroll down a bit to type a shortcut name.**

7. **Choose an icon for the shortcut.**

8. **Touch the Save button.**

   The Navigation shortcut is placed on the Home screen.

To use the shortcut, simply touch it on the Home screen. Instantly, the Maps app starts and enters Navigation mode, steering you from wherever you are to the location referenced by the shortcut.

> ✓ I keep a Navigation shortcut to my home on the Home screen. It helps me quickly find my way back to my house when I'm fleeing the authorities.
>
> ✓ See Chapter 22 for additional information on creating Home screen shortcuts.
>
> ✓ I keep all Navigation shortcuts in one place, on the last Home screen to the right.

Where Are Your Friends?

A feature in the Maps app is Latitude. It's a layer you can apply to a map, but it's also the name of its own app, Latitude.

The *Latitude* social networking service lets you share your physical location with your friends, also assumed to be using Latitude. Being able to more easily know where your friends are makes it possible to meet up with them — or, I suppose, to avoid them. It's all up to you.

To join Latitude, you press the Menu soft button when viewing a map and then choose the Latitude command. Or, you can just start the Latitude app.

After opening Latitude, read the information and then touch the Allow & Share button to continue. If you don't see the Join Latitude command, you've already joined; start Latitude by choosing the Latitude command.

To make Latitude work, you add friends to Latitude and those friends need to use Latitude. After adding Latitude friends, you can share your location with them as well as view their locations on a map. You can also chat with Google Talk, send them email, get directions to their location, and do other interesting things.

To disable Latitude, press the Menu soft button when Latitude is active and choose the Settings command. Choose the option Sign Out of Latitude or Turn Off Latitude. (You may have to scroll the list of commands to find it.)

Capture the Moment

*P*roving once again that the modern cell phone is the equivalent of the Swiss army knife, another amazing ability of your ThunderBolt phone is that it can shoot both still images and video. I'd normally prattle on about how silly it is to use a phone as a camera. As my photography buddies scoff, "Real cameras don't have ringtones." Yet I must confess that the ThunderBolt features a very nice camera. Sure, it's not a Nikon, but just try to make a birthday greeting phone call with a CoolPix — not gonna happen.

Say "Cheese"

The notion behind saying "Cheese" whenever you take a picture is to subtly force everyone to smile. The long-E sound in *cheese* kind of does that. Even so, the phrase has become so popular in other cultures that you'll find people saying their word for *cheese* whenever a picture is taken. They don't even find it important for their word for *cheese* to have the long-E sound. No, apparently the only important thing is to mention a cultured dairy product when you have your picture taken.

Taking a picture

Of the two cameras on the ThunderBolt, the heavy-duty one is on the phone's rump. It's the 8-megapixel (MP) camera, which also has a flash. The front-facing camera is only 1.3MP and has no flash, so most of the time you're taking a picture, you hold the phone about a foot from your face and look at the touchscreen.

To begin your photography adventure on the phone, start the Camera app, found on the All Apps screen and also on the first Home screen to the left of the main Home screen. After starting the Camera app, you see the main Camera screen, as illustrated in Figure 14-1.

To take a picture, point the camera at the subject and touch the Shutter button, shown in Figure 14-1.

After you touch the Shutter button, the ThunderBolt may whir as you hear it focus, and then you may hear a mechanical shutter sound play. Finally, the image you just took appears briefly on the screen for your review, as shown in Figure 14-2.

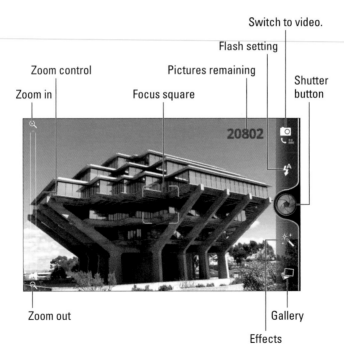

Switch to video.

Flash setting

Zoom control   Pictures remaining

Zoom in   Focus square   Shutter button

20802

Zoom out   Gallery

Effects

Figure 14-1: Camera mode.

Set As   Share |Camera App

Delete image.

Image preview

Figure 14-2: Reviewing a picture you just shot.

You can wait a few seconds to return to Camera mode and take more pictures, or use the buttons referenced in Figure 14-2 to work with the picture. You have to be quick! If the image goes away, you can touch the Gallery button, shown in Figure 14-1, to see the photo again.

- As on other digital cameras, the shutter doesn't always snap instantly when you shoot the picture; the camera takes a moment to focus, the flash may go off, and then you hear the shutter sound effect.

- The ThunderBolt can be used as a camera in either landscape or portrait orientation. The icons on the buttons rotate individually as you switch between orientations.

- If your pictures appear blurry, ensure that the camera lens on the back of the phone isn't dirty.

✔ The pictures you take with your ThunderBolt are saved to the MicroSD card. They remain on the MicroSD card until you delete them.

✔ Reviewing and organizing all the pictures on your ThunderBolt is the job of the Gallery app. See Chapter 15 for information on the Gallery app.

✔ Chapter 15 also offers specifics on how to use the Share command (see Figure 14-2).

✔ Use the Zoom control, illustrated in Figure 14-1, to digitally zoom in and out as you frame a picture. Don't forget to zoom back out!

✔ Touch the screen to move the focus square (refer to Figure 14-1). The ThunderBolt camera uses whatever item is in the square as the camera's focus plane.

✔ Use the Set As button (refer to Figure 14-2) to quickly assign to a contact the image you just took, make the image the phone's wallpaper (Home screen background), or save the image and location in the Footprints app.

✔ You can take as many pictures with your ThunderBolt as you like. The number of pictures you have available appears briefly as you start the Camera app. Look for the number next to the Switch button, illustrated in Figure 14-1. The number also appears right after you take a shot.

✔ Though you can delete an image right after you take it, I recommend that you do your deleting and other photo management duties by using the Gallery app, discussed in Chapter 15.

✔ The next time you're face-to-face with a contact, remember to snap the person's photo.

✔ Use the Effects button to apply a photographic enhancement to your photo. Previews of the effects appear on the same edge of the screen as the Zoom control, shown in Figure 14-1.

✔ The ThunderBolt stores pictures in the JPEG image file format. Images are stored in the DCIM/100MEDIA folder on the MicroSD card; they have the JPG filename extension.

Setting the flash

You can use one of three settings for the flash when taking still shots on your ThunderBolt, as described in Table 14-1.

Table 14-1		Camera Flash Settings on the ThunderBolt
Setting	*Icon*	*What the Flash Does*
Auto		Activates during low-light situations but not when it's bright out
On		Always activates
Off		Never activates, even in low-light situations

To change or check the flash setting, touch the Flash Setting button on the Camera app screen, as shown in Figure 14-1. The icon that appears reflects the current flash state.

✔ A good time to turn on the flash is when you're taking pictures of people or objects in front of something bright, such as Aunt Carol holding her prize-winning peach cobbler in front of a burning building.

✔ Only the rear camera features a flash.

Changing the resolution

The ThunderBolt's rear camera has several resolutions at which you can take an image. To set the resolution, follow these steps before you snap the picture:

1. **Press the Menu soft button.**

2. **Choose Resolution.**

3. **Select a resolution from the list.**

4. **Press the Down button, or just wait a few moments, to return to the Camera app.**

    The Down button is shown in the margin.

The only way to determine the camera's current resolution setting is to work through Steps 1 and 2 and peruse which setting is active.

✔ There's no reason to take at their highest resolution the images you intend to upload to the Internet. Choosing the Small or 1M resolution is fine for social networking and web page images.

✔ The smaller the resolution, the more images you can store inside the phone (on the MicroSD card).

✔ You cannot change the resolution when taking a self-portrait. See the next section.

✔ Resolution settings control the image quality. Technically, the higher the resolution, the more information is in the picture. This concept comes into play when you choose to print an image: Higher resolutions print better. When you plan to take images only for email or posting on the Internet, lower resolutions are fine.

Taking a self-portrait

Who needs to pay all that money for a mirror when you have the ThunderBolt? Not only is the front-facing camera useful for video chat, but you can also take a picture of yourself, alone or with others, by simply following these steps:

1. **Start the Camera app.**

2. **Press the Menu soft button.**

3. **Choose Switch Camera.**

   When you see yourself, you know that you've chosen properly.

The rear camera is known as the *main* camera. The front-facing camera is known as the *secondary* camera.

Setting the location

The ThunderBolt not only takes a picture but also keeps track of where you're located on Planet Earth when you took the picture — as long as you've turned on that option. The feature is Geo-Tag, and here's how to ensure that it's on:

1. **While using the Camera app, press the Menu soft button.**

2. **Scroll the list of commands until you find the item titled Geo-Tag Photos.**

3. **Ensure that a green check mark is in the box next to Geo-Tag Photos.**

   If not, touch the gray box to put a check mark there.

Not everyone is comfortable with the phone recording a picture's location, so you can turn off the option. Just repeat these steps, but in Step 3 remove the green check mark by touching the box.

See Chapter 15 for information on reviewing a photograph's location.

Adjusting the camera

Plenty of interesting settings are in the Camera app. All the settings are accessed from the menu that pops up when you press the Menu soft button while using the Camera app.

Rather than bore you by describing all the settings, here are some of my favorites:

Self-Timer: Touch to activate a delay after you touch the shutter button, such as when you set up the phone's camera and then want to rush around and get yourself into the picture. Options are Off for no self-timer, and 2 seconds or 10 seconds for Self-Timer mode.

Review Duration: Set the number of seconds that the Camera app displays the images you just took. Options are No Review, which disables the review; 2- and 5-second review times; and No Limit, which sets an unlimited review period. Touch the Back soft button to take the picture again when the No Limit option is set.

Widescreen: Choose this option to switch between a widescreen aspect ratio (5:3) and the standard photographic aspect ratio (4:3). The *aspect ratio* is the difference between a picture width and its height.

Shutter Sound: Place a check mark by this item to have the Camera app play a shutter sound when you snap a picture. Remove the check mark to disable this feature.

Reset to Default: Choose this item to change all Camera app options to their original settings. This item is a good choice to make when you've goofed things up so badly that you don't know which options to reset to return things to "normal."

Moving Pictures

No matter how hard you try, you can never snap still images fast enough to capture all the action. There's no need to worry about acquiring quick-button skills on the ThunderBolt touchscreen, however. That's because you have the Camcorder app, which lets you capture moving images just as easily as the phone takes still pictures.

Capturing video

The video capturing abilities of your ThunderBolt phone are accessed by using the Camcorder app. This app is found, like all other apps on the phone, on the All Apps screen.

I'll let you in on a secret: The Camcorder app is the same app as the Camera app, covered earlier in this chapter. It's simply a different mode for that app. In fact, you can easily switch between both apps: The Switch button is illustrated earlier, in Figure 14-1, which shows the Camera app, and in Figure 14-3, which shows the interface for the Camcorder app. See? Same difference.

Start shooting the video by pressing the Record button, shown in Figure 14-3. The screen doesn't change much, though the video's duration appears by the Record button, which becomes the Stop button.

Touch the Stop button when you're done recording video. A preview screen appears for about two seconds, as shown in Figure 14-4. To review your video, touch the Play button, shown in Figure 14-4.

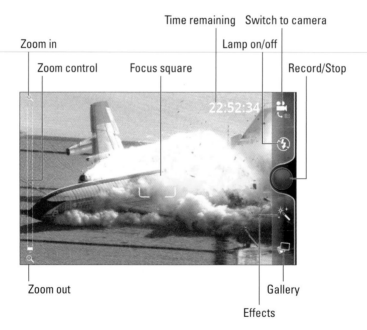

Figure 14-3: Video mode on the ThunderBolt.

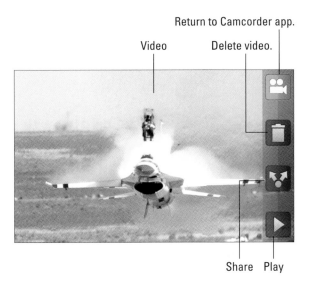

Figure 14-4: Reviewing a video.

✔ Hold the ThunderBolt steady! The camera still works when you whip around the phone, but wild gyrations render the video unwatchable.

✔ The length of video you can record is limited by how much storage space is available on the MicroSD card.

✔ To peruse, as well as manage, your videos, use the Gallery app. See Chapter 15.

✔ Adjust the focus while shooting by touching the screen. A focus square appears at the spot where you touch, which tells the camera where to focus.

✔ If you need more light, touch the Flash button on the screen, which turns on the camera's LED lamps. Touch again to turn off the lamps.

✔ Add video filters and effects by pressing the Effects button and choosing an effect from the scrolling list. For example, to shoot in "black and white," touch the Effects button and choose the Grayscale option. To remove effects, choose No Effect from the scrolling list.

✔ To use the front-facing camera for view recording, press the Menu soft button and choose the Switch Camera command. The front-facing camera is the *secondary* camera; the rear camera is the *main* camera.

✔ The best way to review, manage, and delete the videos you've shot is by using the Gallery app. See Chapter 15.

✔ Also see Chapter 15 for information on publishing your video to YouTube.

✔ The Camera app stores your videos on the MicroSD card in the DCIM/100MEDIA folder. The videos are saved in the 3GPP video file format, which is similar to the MP4 file format. Video files have the .3gp filename extension.

Changing video quality

Though it may seem that choosing high quality or high definition (HD) all the time is the best option, that's not always the case. For example, video you shoot for YouTube need not be of HD quality. Multimedia text messaging (known as *MMS*) video should be of very low quality or else the video won't attach to the message. Also, HD video uses a heck of a lot more storage space on the ThunderBolt's MicroSD card.

To set the video quality while using the Camcorder app, press the Menu soft button and choose the Video Quality command. Here's a rundown of the options and my recommendations for using them:

HD 720P (1280 x 720): The highest-quality setting is best suited for video you plan to show on a large-format TV or computer monitor. It's useful for video editing or for showing important events, the local Eyewitness News, or alien invasions, for example.

Widescreen (800 x 480): This option has a good quality for shooting video when you don't know where the video will end up. This setting is the one the Camcorder app uses automatically.

High (640 x 480): This setting, good for quality Internet video, doesn't enlarge well.

Low (320 x 240): This one is a good choice for medium-quality YouTube and web videos. The files are small and load quickly over an Internet connection. This setting isn't good for viewing videos in a larger format.

MMS (176 x 144): This setting is designed for use with text messaging video attachments. See Chapter 9 for more information on MMS.

Online (HD, 10 minutes): This setting is best for sharing HD-quality video online and is well suited for HD YouTube uploads.

Online (High, 10 minutes): This second option for sharing videos online lacks HD quality.

Check the video quality *before* you shoot! Especially if you know where the video will end up (on the Internet, on a TV, or in an MMS), it helps to set the quality first.

15

Images and Videos in Your Phone

A nerd would tell you that all those pictures and videos you shoot using your ThunderBolt are stored in a compressed digital format, electronically encoded on the phone's removable media card. This description is technically correct, but it doesn't do you any good.

It's true that all the pictures and videos you shoot with the ThunderBolt are stored in the phone. They're kept in a digital photo album, called the Gallery. In the Gallery, you find all the pictures and videos you've taken as well as other images and videos that were copied to the phone or synchronized with various online accounts.

Your Phone's Gallery

The nerd is correct: Images and videos on your phone are stored in a compressed digital format. They exist as a series of ones and zeroes, all of which means nothing unless you have a way to interpret this binary information into something you can see. The app on your phone that handles this job is the *Gallery*.

Using the Gallery

To view images in your phone, start the Gallery app by choosing its icon from the All Apps screen. When the Gallery app opens, you see your visual media (pictures and videos) organized into albums, as shown in Figure 15-1.

The number and variety of albums depend on how you synchronize your phone with your computer, which apps you use for collecting media, or which photo sharing services you use on the Internet, such as Flickr or Picasa.

Touch a folder to open it and view thumbnails of the pictures or videos it contains. The media is displayed in a grid when the phone is held vertically; when you hold the phone horizontally, the images appear one after the other and the thumbnails are larger.

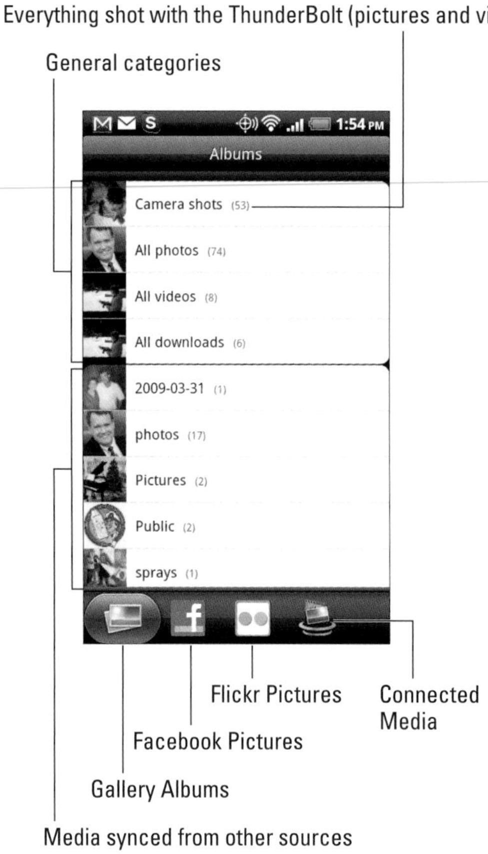

Everything shot with the ThunderBolt (pictures and videos)

General categories

Flickr Pictures   Connected Media

Facebook Pictures

Gallery Albums

Media synced from other sources

Figure 15-1: Albums in the Gallery.

Videos stored in a folder appear with the Play button (shown in the margin) on their thumbnails.

Figure 15-2 illustrates how images look when you open a folder and the ThunderBolt is held in a vertical orientation. Generally speaking, the images are organized by date, oldest first (top or left).

To view a single image or a video, touch it with your finger. The image appears in full size on the screen, similar to the one shown in Figure 15-3. You can rotate the phone horizontally (or vertically) to see the image in another orientation — for example, to view a full-screen portrait or landscape photo.

Later sections describe in more detail what you can do when viewing an image, as shown in Figure 15-3.

Videos play when you choose them. The Video Playing screen is illustrated in Figure 15-4. To see the controls (illustrated in Figure 15-4), touch the screen while the video is playing.

Figure 15-2: Looking at an album.

Album

Picture number | Total items in album

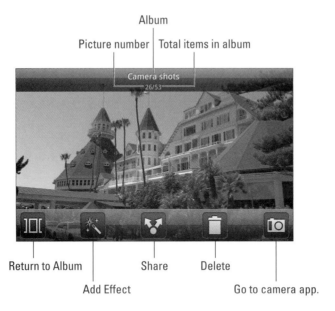

Return to Album   Share   Delete

Add Effect   Go to camera app.

Figure 15-3: Examining an image.

Video

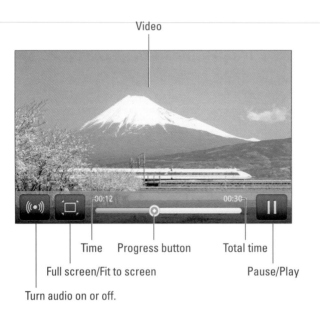

Time   Progress button   Total time

Full screen/Fit to screen   Pause/Play

Turn audio on or off.

Figure 15-4: Watching one of your videos.

To return to the album, press the Back soft button or press the Album button, shown in Figure 15-3. The Gallery button shown in Figure 15-2 takes you back from an album to the main Gallery screen.

✒ You can use the buttons on the tab at the bottom of the Gallery's main screen (refer to Figure 15-1) to access photos on Facebook or Flickr or to connect to a media server to view even more pictures and videos.

✒ Videos in the Gallery play in one orientation only.

✒ Touch a video or an image to redisplay the onscreen menu, as shown in Figures 15-3 and 15-4.

✒ The Video app, found on the All Apps screen, provides a shortcut to the Gallery, one that shows only videos.

✒ To keep the Gallery handy, considering placing its shortcut icon on the Home screen. Refer to Chapter 24.

✒ Refer to Chapter 11 for information on downloading photos from the web.

✒ See Chapter 20 for information on the doubleTwist program, which can be used to synchronize images and videos between the ThunderBolt and your computer.

✒ If you've configured a computer or game machine for streaming media sharing, you can touch the Connected Media button in the Gallery (refer to Figure 15-1) to access media on your PC or Xbox or a similar gizmo. That way, you can view pictures and videos and listen to music from that device on your ThunderBolt over the Wi-Fi connection. Yeah, it's complex, which is why I've flagged this point with a nerd-guy icon.

Finding a picture's location on the map

In addition to snapping a picture, the ThunderBolt can save the location where you took the picture. This information is obtained from the phone's GPS, the same tool used to find your location on a map. In fact, you can use the information saved with a picture to see exactly where the picture was taken.

For example, Figure 15-5 shows the location where I took the image shown in Figure 14-1 (from Chapter 14). The location was saved by using GPS technology and is available as part of the picture's data.

To see where you've taken a picture, follow these steps:

1. **Summon the image in the Gallery.**

2. **Press the Menu soft button.**

3. **Choose Show on Map.**

   The spot where you took the picture appears in the Maps app.

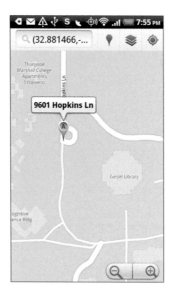

Figure 15-5: A picture's location.

Not every image has location information. In many cases, the ThunderBolt might be unable to acquire a GPS signal. When that happens, location information is unavailable.

- ✔ As far as I can tell, videos don't store location information on the ThunderBolt.
- ✔ The phone's GPS can be disabled for images you capture. Refer to Chapter 13 for information on how to turn GPS on or off when taking pictures.

Watching a slide show

One key command available in the Gallery app is Slideshow, which allows you to view the photos in an album one after the other — just like those slide shows Uncle Lloyd bored you with when you were a kid. This time, however, it's *you* showing the slides.

To begin your slide show adventure, obey these steps:

1. **Open the album.**

   Slide shows comprise only a single album.

2. **Press the Menu soft button.**

3. **Choose the Slideshow command.**

   The slide show begins and your audience drifts happily to sleep.

You can touch the screen during the slide show to see onscreen controls, as shown in Figure 15-6.

✔ Slide show settings are made by pressing the Menu soft button while viewing the slide show and then touching the Settings command.

✔ Videos aren't part of the Gallery app's slide show.

✔ If the screen dims during a slide show, it's simply the phone's screen saver kicking in. Touch the screen to restore normal brightness.

✔ The ThunderBolt has a command to attach a slide show to a multimedia text message (MMS), which works differently from the Gallery's app's slide show. See Chapter 9 for more information.

Back to Album   Previous Image   Pause/Play   Next Image

Figure 15-6: Onscreen slide show controls.

Assigning an image to a contact

You can set any image for a contact. It doesn't necessarily have to be an image you've taken with the ThunderBolt; any image you find in the Gallery will do. Follow these steps:

1. **View an image in the Gallery.**

2. **Press the Menu soft button.**

3. **Choose Set As.**

   If the Set As command doesn't appear, you cannot set that image for a contact; not every album allows its images to be set for contacts.

4. **Choose Contact Icon.**

5. **If prompted, choose the People app.**

   Other apps may allow you to set contact icons, but in these steps the goal is to set an icon for a contact in the People app.

6. **Scroll the phone's address book and choose a particular contact.**

   You can use the Search command to easily locate a contact when you have an abundance of them.

7. **Crop the image.**

   Refer to the later section "Cropping an image" for detailed instructions on working the crop-thing.

8. **Touch the Save button.**

   Or, touch the Cancel button to chicken out or change your mind.

The contact's image is set and you return to view the image in the Gallery.

Images you set for your Google contacts are instantly synchronized with your Google account on the Internet.

Setting an image as wallpaper

You can choose as the ThunderBolt's Home screen background or wallpaper any picture that's viewable in the Gallery. The steps work like this:

1. **View an image in the Gallery, one that would make a good-looking wallpaper.**

   Wider images work better than portraits. The wallpaper image scrolls slightly as you flick from Home screen to Home screen.

2. **Press the Menu soft button and choose Set As.**

3. **Choose Wallpaper.**

4. **Crop the image.**

   Only the portion in the green rectangle is used for the wallpaper.

5. **Touch the Save button.**

6. **Press the Home button to see the new wallpaper.**

See Chapter 22 for more information on changing the Home screen's wallpaper.

Cropping an image

The Crop command is used to edit an image, slicing out portions you don't want, such as when removing convicts and former spouses from family portraits. To crop an image, obey these directions:

1. **Open the album containing the image.**

2. **Long-press the image you want to crop.**

3. **Choose Edit, and then choose Crop.**

   If the Crop command is unavailable, you have to choose another image. (Not every album lets you modify images.)

4. **Work the crop-thing.**

   Drag the green rectangle around to choose which part of the image to crop. Drag an edge of the rectangle to resize the left and right or top and bottom sides. Use Figure 15-7 as your guide.

Resize top-bottom edges.

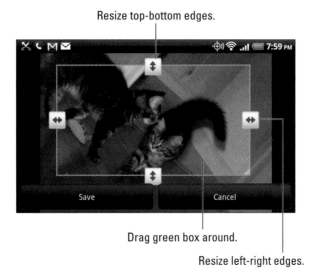

Drag green box around.

Resize left-right edges.

Figure 15-7: Working the crop-thing.

5. **Touch the Save button when you're done cropping.**

   Only the portion of the image within the green rectangle is saved; the rest is discarded.

There's no way to undo a crop action after you've touched the Save button.

Working the green rectangle to crop an image can be frustrating. When the image zooms in too small, it's difficult to zoom back out. At that point, you can touch the Cancel button and start over again.

Rotating pictures

Which way is up? Well, the answer depends on your situation. For taking pictures with the ThunderBolt, sometimes images just don't appear "up," no matter how you turn the phone. To fix the situation, heed these steps:

1. **Open the album containing the image in need of rotation.**
2. **Long-press the cockeyed image.**
3. **Choose the Edit command.**
4. **Choose Rotate Left or Rotate Right to spin the image 90 degrees counterclockwise or clockwise, respectively.**

Some images can't be rotated, such as images synced from online sites. In that case, go to the online site to edit the image.

Videos cannot be rotated.

Deleting pictures and videos

It's entirely possible, and often desirable, to remove unwanted, embarrassing, or questionably legal images and videos from the Gallery. To zap a single image or video, follow these steps:

1. **Summon the album containing image or video you want to get rid of.**
2. **Long-press the image or video.**
3. **Choose Delete.**
4. **Choose OK to confirm.**

   Or, touch Cancel to chicken out.

For mass deletion, gingerly obey these steps:

1. **Open the album.**
2. **Touch the Delete button.**

   Refer to Figure 15-2 for its location; it looks like a trash can.

3. **Touch every item you want to delete.**

   A red X appears over the thumbnail of whichever item you touch. It's the item's marked-for-death symbol.

   To remove the red X, touch the item again.

4. **Touch the Delete button to instantly zap items marked for death.**

You cannot undelete an image or a video you've deleted. There's no way to recover an image using available tools on the ThunderBolt.

Some images and videos cannot be deleted. Specifically, you cannot remove pictures shared from your photo-sharing websites. In that case, visit the website itself to remove the image.

Set Free Your Pics and Vids

My grandmother used to carry several sets of wallet-size photo albums with her everywhere. I'm sure her friends looked forward to those moments when she'd pull out the album and show off us grandkids to impress her friends.

Today's grandmother doesn't need various wallet-size photo albums. Instead, modern grandparents — as well as the rest of us — tote around our phones. Showing someone a picture on a cell phone has become common these days. What's better than that is sharing these images electronically on the Internet, over email, on social networking sites, and in many other ways, as covered in this section.

Printing and sharing with Bluetooth

Bluetooth is perhaps the most complex way to share files, mostly because its setup is complex and file transfer works more slowly than it does over a USB connection. In fact, if you've never transferred files by using Bluetooth, I recommend that you simply use a USB direct connection, which is covered in Chapter 20.

Now that you've survived reading the preceding paragraph, *Bluetooth* is a simple and quick way to send files to Bluetooth-equipped computers. If you have a Bluetooth printer, you can send a picture from your ThunderBolt to the Bluetooth printer, in which case it prints.

Generally speaking, the Bluetooth picture-sharing operation works like this:

1. **Pair your phone with the Bluetooth device.**

   Detailed instructions on pairing are found in Chapter 19.

2. **In the Gallery app, open the album containing the image you want to send or print.**

3. **Touch the Share button.**

   Refer to Figure 15-2.

4. **Choose Bluetooth from the Share Via menu.**

5. **Touch in the album any images you want to share.**

6. **Touch the Next button.**

7. **Choose the paired device from the list of Bluetooth devices.**

8. **Obey whatever directions appear on the device to accept the images.**

   For example, on a Bluetooth-equipped PC, a pop-up window or a notification icon appears. Choose whichever option accepts the files from your ThunderBolt and transfers them to your PC.

 If the upload is successful, you see the Upload notification icon appear, as shown in the margin. For a Bluetooth printer, the image prints. For a computer, you have to hunt down where the file was saved. On my PC, which uses a Belkin Bluetooth adapter, files sent from the ThunderBolt appear in the folder named `Documents\Bluetooth Exchange Folder`.

   ✔ PCs don't generally come with Bluetooth adapters installed. You can obtain USB Bluetooth adapters at any computer or office supply store.

   ✔ Most newer Macs come equipped with Bluetooth. Use the Bluetooth icon on the menu bar to manage Bluetooth connections and send files to the ThunderBolt.

   ✔ Bluetooth printers show the Bluetooth symbol somewhere on the printer, as shown in the margin.

Sharing with the Share menu

The Share command is used in the Gallery to free your images and videos from the confines of your phone.

You can either share an individual image by long-pressing it and choosing the Share command or share a group of items by touching the Share button (refer to Figure 15-2), choosing a sharing method, and then tagging multiple items to share.

The menu that appears when you choose the Share command contains various options for sharing media, similar to the one shown in Figure 15-8. You may see more or fewer items on the Share menu, depending on which software you have installed on your phone, which Internet services you belong to, and which type of media is being shared.

 When viewing a video, pause the video and press the Menu soft button. Choose the Share command to see a menu similar to the one shown in Figure 15-8.

The following sections describe some of the items you can choose from the menu and how the media is shared.

Email sharing (Gmail and Mail)

When you share media with your email program, you're creating an email attachment. After choosing pictures or videos, simply create your email as you normally would: Fill in the To, Subject, and Message text boxes as necessary. Touch the Send button to send the media as an attachment.

Figure 15-8: Sharing options for media.

- ✔ See Chapter 10 for more information on Email and Gmail on the ThunderBolt.

- ✔ You may not be able to send video files as email attachments. That's probably because some video files are humongous. They would not only take too long to send but also might be too big for the recipient's inbox.

- ✔ As an alternative to sending large video files, consider uploading them to YouTube instead. See the later section "Posting your video to YouTube."

Facebook

To upload a mobile image to Facebook, choose a Facebook command from the Share menu. You might see more than one Facebook command, depending on whether you've installed the specific Facebook app.

- ✔ The Facebook for HTC Sense command uses the Friend Stream app to upload your picture to Facebook.

- ✔ The Facebook command uses the Facebook app.

- ✔ Other Facebook and social networking apps may sport additional commands on the Share menu (refer to Figure 15-8).

Both the Friend Stream and Facebook apps are covered in Chapter 12.

Optionally, add a caption to your image and then touch the Upload button to send it to Facebook.

Flickr

To send one or more images to your Flickr account, choose the Flickr option from the Share Via menu. Follow these steps to complete the process:

1. **Optionally, rename the image.**

   The image's original filename appears above the preview picture. Those filenames are obtuse, so feel free to touch the white text box and give your photo a better name.

2. **Optionally, touch the image to add a caption or description.**

   You must specifically touch the text *Tap to Add Description,* which is a text box in which you can type.

3. **Optionally, touch the Tag This Photo command to add descriptive tidbits of text.**

   These *tags,* or tidbits of text, help others find the image. For example, you can specify as tags the names of people in the picture, the date, the location, or the event.

4. **Touch the Upload button.**

   The button lied! There's more to do:

5. **Optionally, choose an album to post the image to.**

6. **Optionally, set the picture's visibility.**

7. **Touch the Done button to send the image off to Flickr.**

Images appear nearly instantaneously on your Flickr account, which can be viewed on the Internet (at www.flickr.com) or in the Gallery app by touching the Flickr button at the bottom of the main screen, illustrated in Figure 15-1.

Messages

Media can be attached to a text message, which then becomes the famous multimedia message, or *MMS,* that I write about in Chapter 9. In fact, the Messages item on the Share Via menu is simply a shortcut to attach media to a message, as described in Chapter 9.

After choosing the Messages sharing option, you may see a notice that the image is being compressed. That's okay; you can't attach large images to a text message.

On the ThunderBolt, images are attached to text messages as a slide show, even if you select only one image. That's okay: Touch the Done button and then compose your text message as you normally would. Touch the Send button to send the message.

Peep

Images are shared on the popular Twitter social networking site by saving the image on a photo host website and then tweeting the image's link. For the Peep app, the Twitgoo (www.twitgoo.com) image sharing site is used (though it can be changed from within the Peep app).

After you choose the image and choose Peep, the image is uploaded and its link is included in a tweet. You can add more text, or just touch the Update button to share the link, and the image, with your fellow twits.

YouTube

The YouTube sharing option appears when you've chosen to share a video from the Gallery. See the next section.

Grabbing pics and vids with a media manager

One handy way to get media out of your phone and into your computer is to use a media manager on the computer. This type of program is used to gather photos and videos from digital cameras, pulling them out of the camera and organizing them on your computer.

For example, the Windows Photo Gallery app can be used to access the ThunderBolt's digital media, not only copying images from the phone but also sending some of your favorites over to the phone. Similar programs exist for both the PC and the Mac.

The key to getting media management software to synchronize with the ThunderBolt is to make a USB cable connection between your phone and your computer. See Chapter 20 for the details on this connection.

Posting your video to YouTube

The best way to share your video creation with the known universe is to upload it to YouTube. As a Google account holder, you also have a YouTube account. You can use the YouTube app on the ThunderBolt along with your account to upload your videos to the Internet, where everyone can see them and make rude comments about them. Here's how:

1. **If you can't get a 4G LTE signal, ensure that the Wi-Fi connection is activated.**

   You might still see a warning about using Wi-Fi rather than the digital cellular connection, even if you have a 4G LTE connection. See Chapter 19 for information on how to turn on the Wi-Fi connection.

2. **Start the Gallery app.**

3. **Open the album containing the video you want to upload.**

4. **Long-press the video's thumbnail.**

5. **Choose the Share command.**

6. **Choose YouTube.**

7. **Type the video's title.**

8. **Touch the More Details button.**

9. **Optionally, type a description, specify whether to make the video public or private, add tags, or change other settings.**

10. **Touch the Upload button.**

    You return to the Gallery and the video is uploaded. It continues to upload, even if the ThunderBolt falls asleep.

To view your video, open the YouTube app. It's found on the All Apps screen. Press the Menu soft button and choose the My Channel command. If you don't see your recently uploaded video in the Uploads list, it's probably still processing

See Chapter 17 for more information on using YouTube.

Tunes on the Go

During one fad back in the 1970s, just when touch-tone phones became popular, you could use the phone's keypad to play "music." Granted, the range of tunes was rather limited. Even so, you could dial various numbers to play such popular ditties as *Camptown Races* and *Yankee Doodle* and even a few classical tunes. It was a fad.

You don't have to worry about making music on your ThunderBolt by dialing various phone numbers. That's because the phone comes with a complete music-playing app. By using this app, the ThunderBolt becomes your portable MP3 music player. You can buy tunes directly on the phone, import them from your computer, and play them wherever you are.

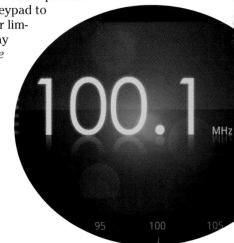

Music, Music, Music

John Cage promoted the idea that anything can be music — even 4 minutes and 33 seconds of silence. I'm sure he'd have a wonderful time showing how the ThunderBolt can make music even without turning on the device. You're probably not that *avant garde*. Instead, you'll probably just grab yourself a nice pair of earphones or a headset and run the Music app, as covered in this section.

Browsing your music library

The music-playing duties of your ThunderBolt are deftly handled by the app coincidentally named Music. The app can be found on the All Apps screen.

The Music app is shown in Figure 16-1. You may not see any music on your ThunderBolt just yet. That's okay; the later section "Get Some Music into Your Phone" explains how to add music.

Albums stored on the phone

Boring, generic artwork

Album artwork

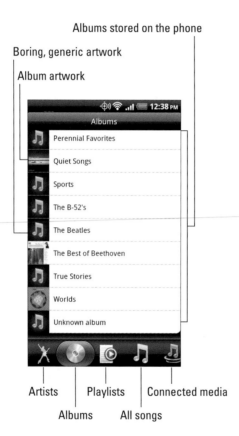

Artists   Playlists   Connected media

Albums   All songs

Figure 16-1: The Music library.

The music stored on your ThunderBolt is presented in the Music app by category. Figure 16-1 shows the Albums category. Here are all five categories:

Artists: Music is listed by recording artist or group.

Albums: Music is organized by album. The albums are created when music is purchased or imported to the ThunderBolt.

Playlists: Music is organized into playlists that you create.

All Songs: All music (songs and audio) is listed individually and in alphabetical order.

Connected Media: Access computers or other devices that can share music over the Wi-Fi network.

These categories are merely ways the music is organized — ways to make the music easier to find when you may know an artist's name but not an album title. Well, except for the Connected Media item, which involves a lot of network setup on a PC or game console to get it to work. That ugly topic is covered in Chapter 20.

A *playlist* is a list you create yourself to organize songs by favorite, theme, or mood or whatever other characteristic you want. The section "Music Organization," later in this chapter, discusses playlists.

- Music is stored on the ThunderBolt's MicroSD card.

- The size of the MicroSD card limits the total amount of music that can be stored on your phone. Also, consider that storing pictures and videos horns in on some of the space that can be used to store music.

- Two types of album artwork are used by the Music app. Some music may have the original album artwork, as shown in Figure 16-1. Most music, lamentably, features the generic album cover.

- There's no easy or obvious way to apply album cover artwork to music with a generic album cover.

- When the ThunderBolt can't categorize music, it uses the word *Unknown,* as in Unknown Artist or Unknown Album. This label applies to music you copy manually to your phone, but it can also apply to audio recordings you make yourself. Music you purchase, or import or synchronize with a computer, generally retains its artist and album information. (Well, the information is retained as long as it was supplied on the original source.)

Listening to a tune

To listen to music by locating a song in your music library, as described in the preceding section. Locate songs by choosing a category at the bottom of the Music app and choose an artist, an album, or a playlist, if necessary. Touch the song title, and the song plays, as shown in Figure 16-2.

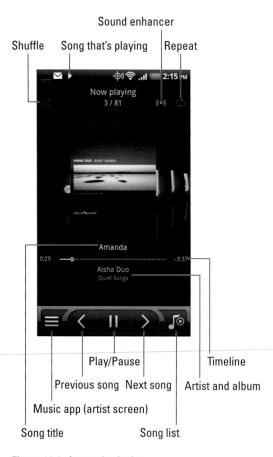

Sound enhancer

Shuffle   Song that's playing   Repeat

Now playing
3 / 81

Amanda

0:29   -3:37

Aisha Duo
Quiet Songs

Play/Pause

Previous song Next song   Artist and album

Music app (artist screen)

Song title   Song list

Timeline

Figure 16-2: A song is playing.

While the song plays, you're free to do anything else on the ThunderBolt. In fact, the song continues to play even when the phone goes to sleep.

After the song is done playing, the next song in the list plays. Touch the Song List button (refer to Figure 16-2) to review the songs in the list.

The next song doesn't play if you have the Shuffle button activated (refer to Figure 16-2). In that case, the Music app randomly chooses another song from the list. Who knows which one is next?

The next song also might not play when you have the Repeat option on: The three Repeat settings are illustrated in Table 16-1, along with the Shuffle settings. To change settings, simply touch either the Shuffle or Repeat button.

Table 16-1	Shuffle and Repeat Button Icons	
Icon	*Setting*	*What Happens When You Touch the Icon*
	No Shuffle	Songs play one after the other
	Shuffle	Songs are played in random order
	Repeat Off	Song don't repeat
	Repeat All	All songs in the list play over and over
	Repeat Current	The same song plays over and over

To stop the song from playing, touch the Pause button (refer to Figure 16-2).

 A notification icon appears while music is playing on the ThunderBolt, as shown in the margin. To quickly summon the Music app and see which song is playing, or to pause the song, pull down the notifications and choose the first item, which is the name of the song that's playing.

The song title also appears on the phone's unlock screen. In fact, for a few moments, controls appear on the lock screen that let you repeat the song, pause or play, or fast-forward to the next song. If the controls go away, touch the lock screen to see them again.

- ✔ Most of the controls you see in Figure 16-2 aren't available when you hold the phone in a horizontal location.

- ✔ The volume can always be set by using the Volume switch on the side of the ThunderBolt: Up is louder, down is quieter.

- ✔ Determining which song plays next depends on how you chose the song that's playing. If you choose a song by artist, all songs from that artist play, one after the other. When you choose a song by album, that album plays. Choosing a song from the entire song list causes all songs in the Music app's library to play.

✔ When you're browsing your music library, you may see the Speaker icon, similar to the one shown in the margin. This icon flags any song that's playing or paused.

✔ To choose which songs play after each other, create a playlist. See the section "Music Organization," later in this chapter.

✔ After the last song in the list plays, the Music app stops playing songs — unless you have the Repeat All option set, in which case the list plays again.

✔ You can use the ThunderBolt's search abilities to help locate tunes in your music library. You can search by artist name, song title, or album title. The key is to press the Search soft button when you're using the Music app. Type all or part of the text you're searching for and touch the Search button on the onscreen keyboard. Choose the song you want to hear from the list that's displayed.

Being the life of the party

You need to do four things to make your ThunderBolt the soul of your next shindig or soirée:

✔ Connect it to a stereo.

✔ Use the Shuffle command.

✔ Set the Repeat command.

✔ Provide plenty of drinks and snacks.

Hook the ThunderBolt into any stereo that has a standard line input. You need, on one end, an audio cable that has a mini-headphone jack and, on the other end, an audio input that matches your stereo. Look for this type of cable at Radio Shack or any stereo store.

After your phone is connected, start the Music app and choose the party playlist you've created. If you want the songs to play in random order, touch the Shuffle button.

You might also consider choosing the Repeat All command (see Table 16-1) so that all songs in the playlist repeat.

To play all songs saved on your ThunderBolt, choose the All Songs category and touch the first song in the list.

Enjoy your party, and please drink responsibly.

Get Some Music into Your Phone

Odds are good that your ThunderBolt came with no music preinstalled. It might have: Some resellers may have preinstalled a smattering of tunes, which merely lets you know how out of touch they are musically. Regardless, you can add music to your ThunderBolt in a number of ways, as covered in this section.

Getting music from your computer

Your computer is the equivalent of the 20th century stereo system — a combination tuner, amplifier, and turntable, plus all your records and CDs. If you've already copied your music collection to your computer, or if you use your computer as your main music storage system, you can share that music with your ThunderBolt.

Many music-playing, or *jukebox,* programs are available. On Windows, the most common program is Windows Media Player. You can use this program to synchronize music between your PC and the ThunderBolt. Here's how it works:

1. **Connect the ThunderBolt to your PC.**

   Use the USB cable that comes with the phone.

2. **If necessary, pull down the USB notification.**

   The ThunderBolt may automatically display the Connect to PC screen. If not, pull down the notifications and choose the USB connection-type notification.

3. **Choose Media Sync.**

4. **Touch the Done button.**

   The AutoPlay dialog box appears in Windows, prompting you to choose how best to mount the ThunderBolt into the Windows storage system.

5. **Close the AutoPlay dialog box.**

6. **Start Windows Media Player.**

7. **Click either the Sync Tab or Sync toolbar button.**

   The ThunderBolt appears in the Sync list on the right side of Windows Media Player, as shown in Figure 16-3.

8. **Drag to the sync area the music you want to transfer to your phone (refer to Figure 16-3).**

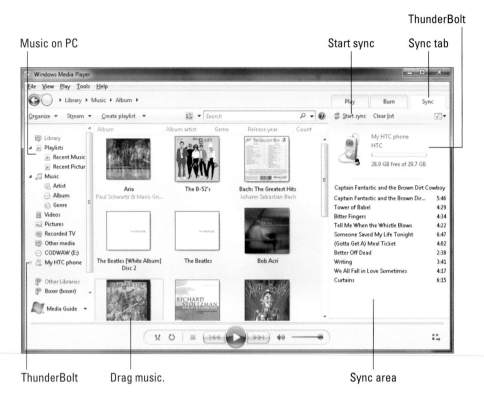

ThunderBolt

Music on PC   Start sync   Sync tab

ThunderBolt   Drag music.   Sync area

Figure 16-3: Windows Media Player meets ThunderBolt.

9. **Click the Start Sync button to transfer the music from your PC to the ThunderBolt.**

   The Sync button may be located atop the list, as shown in Figure 16-3, or found at the bottom.

10. **Close the Windows Media Player when you're done transferring music.**

    Or, you can keep it open — whatever.

11. **On your ThunderBolt, pull down the USB notification and choose the Media Sync item.**

12. **Choose Charge Only from the Connect to PC screen.**

13. **Touch the Done button.**

14. **Unplug the USB cable.**

    Or, you can leave the phone plugged in.

The Rhapsody and TuneWiki apps

Two music apps that ship with the ThunderBolt are Rhapsody and TuneWiki. They provide additional options for obtaining and listening to music on your phone and over the Internet.

Rhapsody is an online music streaming app, though you can also download songs to your phone. For a monthly fee, you gain access to a vast library of tunes you can listen to on your phone, your computer, or any device supported by Rhapsody. The service is so popular that some pundits are blaming Rhapsody for the demise of CD sales.

The TuneWiki all-in-one music program lets you access music already stored on your ThunderBolt, listen to Internet radio, watch music videos, and even do a bit of social networking with music. For example, one of the more interesting things you can do with TuneWiki is see a map of other TuneWiki users near you and peer in on what music they're listening to.

When you have a Macintosh or you detest Windows Media Player, you can use the doubleTwist program to synchronize music between your ThunderBolt and your computer. Refer to the section about synchronizing with doubleTwist in Chapter 20 for more information.

- Chapter 20 contains information about making the USB connection between the ThunderBolt and your computer.

- You must mount the ThunderBolt — specifically, its MicroSD card — into your computer's storage system before you can synchronize music.

- The ThunderBolt can store only so much music! Don't be overzealous when copying your tunes. In Windows Media Player (refer to Figure 16-3), a capacity thermometer-thing shows you how much storage space is used and how much is available on your phone. Pay heed to the indicator!

- Windows Media Player may complain if you try to sync the ThunderBolt to more than one PC. If you do, you're warned after Step 6 in this section. It's not a big issue: Just inform Windows Media Player that you intend to sync with the computer for only this session.

- You cannot use iTunes to synchronize music with the ThunderBolt.

- Okay, I lied in the preceding point: You *can* synchronize music using iTunes but only when you install the *iTunes Agent* program on your PC. You then need to configure the iTunes Agent program to use your ThunderBolt with iTunes. After you do that, iTunes recognizes the ThunderBolt and lets you synchronize your music. Yes, it's technical; hence the icon in the margin.

When the USB connection is on and the ThunderBolt's MicroSD card is mounted into the computer's storage system, you cannot access certain information stored on your phone. That means you cannot play music, look at photos, or access contacts while the MicroSD card is mounted.

Buying music at the Amazon MP3 store

A more specific solution for buying music for the ThunderBolt is to use the Amazon MP3 store. It's a universal approach, which works on just about every Android device, including the ThunderBolt.

The best way to use the Amazon MP3 store is to have an Amazon account. If you don't have one set up, use your computer to visit www.amazon.com and create one. You also need to keep a credit card on file for the account, which makes purchasing music with the ThunderBolt work O so well.

You also need to download the Amazon MP3 app for your phone. Scan the bar code in the margin, and refer to Chapter 18 for information on obtaining new apps.

Follow these steps to buy music from the Amazon MP3 store:

1. **Start the Amazon MP3 app.**

   The Amazon MP3 app connects you with the online Amazon music store, where you can search or browse for tunes to preview and purchase for your ThunderBolt.

2. **If you're prompted to choose between the Store and Cloud Player, choose Store.**

3. **Type some search words, such as an album name, a song title, or an artist name.**

   Or, you can browse the categories to find what you want.

   Your search results appear, if any matches are found, as shown in Figure 16-4.

4. **Touch a result.**

   If the result is an album, you see the contents of the album. Otherwise, a brief audio preview plays.

   When the result is an album, choose a song from the album to hear the preview.

   Touch the song again to stop the preview.

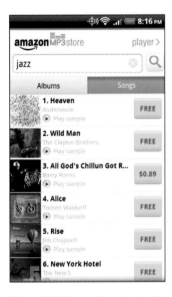

Figure 16-4: Songs found at the Amazon MP3 store.

5. **To purchase the song, touch the big, orange button with the amount in it.**

   For example, a big, orange button in Figure 16-4 specifies *$0.89*. If the button says *Free,* the song is free.

   Touching the button changes the price into the word *BUY*; for free songs, the button changes to *GET*.

6. **Touch the word *BUY* or *GET*.**

7. **If necessary, you may need to accept the license agreement.**

   This step happens the first time you buy something from the Amazon MP3 store.

8. **Sign in to your Amazon.com account: Type your account name or email address and password. Touch the Sign In button.**

9. **Touch the button Save to This Device.**

   Choosing this option saves the song on your ThunderBolt, where it's available using the Music app.

   If you choose to save the song to the Amazon cloud drive, the song is available from any device that can access Amazon's cloud drive, such as another phone, a tablet, or a computer. For this section, however, the goal is to get music into your phone.

10. **Wait while the music downloads.**

Well, actually, you don't have to wait: The music continues to download while you do other things with your ThunderBolt.

No notification icon appears when the song or album has finished downloading. The MP3 Store downloading icon, however, vanishes from the notification part of the screen. It's your clue that the new music is in the phone and ready for your ears.

- ✔ Another bonus with the Amazon MP3 player: the Free Song of the Day and MP3 Daily Deal Album discount. Check in daily for new offers.

- ✔ Amazon emails you a bill for your purchase. It's your purchase record, so I advise you to be a good accountant and print it and then input it into your bookkeeping program or personal finance program at once!

- ✔ Bills are still sent, even for free songs. A bill is a bill.

- ✔ You can review your Amazon MP3 store purchases by pressing the Menu soft button in the Amazon MP3 app and choosing the Downloads command.

- ✔ If possible, try to use Wi-Fi when downloading your purchased music. That way, you don't rack up digital cellular fees.

Music Organization

The Music app categorizes your music by album, artist, song, and so forth, but unless you have only one album and enjoy all the songs on it, that situation probably won't do. To better organize your music, you can create *playlists*. This way, you can hear the music you want to hear, in the order you want, for whatever mood hits you.

Checking your playlists

Any playlists you've already created, or that have been preset on the phone, appear when you choose the Playlists button in the Music app. You see your playlists (if any), similar to the ones shown in Figure 16-5.

To see which songs are in a playlist, touch the playlist name. To play the songs in the playlist, touch the first song in the list.

A playlist is a helpful way to organize music when a song's information may not have been completely imported into the ThunderBolt. For example, if you're like me, you probably have a lot of songs labeled Unknown. A quick way to remedy this situation is to name a playlist after the artist and then add those unknown songs to the playlist. The next section describes how it's done.

Make a new playlist.

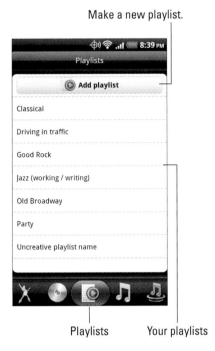

Playlists   Your playlists

Figure 16-5: Playlists in the Music app.

Building a new playlist

Making a new playlist is easy, and adding songs to the playlist is even easier.
Follow these steps:

1. **Touch the Playlist button in the Music app.**

   You see a screen similar to the one shown in Figure 16-5.

2. **Touch the Add Playlist button.**

3. **Type a name for your playlist.**

   Erase the existing, stupid name and type something new. Short and
   descriptive names are best.

4. **Touch the Add Songs to Playlist button.**

   Empty playlists are useless. (Well, unless you're writing a book and
   want to create some playlists with silly names for guffaw value; see
   Figure 16-5). After touching the Add Songs to Playlist button, you see the
   Select Artist screen.

   If you'd rather add songs by album or by music track, choose either the
   Album or All Songs button from the bottom of the screen.

5. **If necessary, open an album or artist entry to see their songs.**

6. **Touch the gray square by a song to select it for inclusion in your playlist.**

   Songs with a green check mark by them are marked for addition to the playlist.

7. **Touch the Add button to assign the songs to your playlist.**

8. **If necessary, repeat Steps 4 through 7 to add even more songs.**

9. **Touch the Save button to create your playlist, teeming with tunes.**

You can have as many playlists as you like on the ThunderBolt and stick as many songs as you like into them. Adding songs to a playlist doesn't noticeably affect the storage capacity of the MicroSD card.

- ✔ You can add songs to a playlist after it's created: Choose the playlist and press the Menu soft button. Choose the Add Songs command to choose more music for the playlist.

- ✔ To remove a song from a playlist, open the playlist and long-press the song. Choose the Delete command and then touch the OK button to confirm.

- ✔ Removing a song from a playlist doesn't delete the song from the ThunderBolt's music library.

- ✔ Songs in a playlist can be rearranged: While viewing the playlist, press the Menu soft button and choose the Change Order command. Use the tab on the far right end of a song title to drag that song up or down in the list. Touch the Done button when you've completed your rearranging duties.

- ✔ To delete a playlist, long-press its name in the list of playlists. Choose the Delete Playlist command. It's gone.

Deleting unwanted music

To purge music from your ThunderBolt, follow these brief, painless steps:

1. **Locate the music that offends you.**

   You can choose any music category but Playlist; deleting music from a playlist doesn't permanently remove it from the phone.

2. **Long-press the musical entry.**

3. **Choose Delete.**

   A warning message appears.

4. Touch the OK button to confirm.

> The music is gone. La, la, la, la: The music is gone.

The music is deleted permanently from the MicroSD card.

By deleting music, you free up storage space, though not really that much. You cannot recover any music you delete. If you want the song back, you have to reinstall it, sync it, or buy it again, as described elsewhere in this chapter.

Phone, Phone, Radio

Are you old enough to remember carrying around a transistor radio? It was once *the* cool accessory. If you long for those halcyon days but prefer not to tote about another gadget, you can rejoice at learning that the ThunderBolt has various radio abilities. It's another way you can use the phone to listen to your favorite tunes.

Turning the ThunderBolt into a cheap FM radio

Nestled on the All Apps screen is the app named FM Radio. It uses mysterious science to pull radio signals from the air and turn them into music for your ears. That's right: You use the FM Radio app to turn the ThunderBolt into a cheap FM radio. It's amazing.

Begin your FM radio journey by plugging a headset into your ThunderBolt. If you don't, the program doesn't even start, and it rudely tells you to plug in a headset.

The headset becomes the phone's FM antenna, which is why it's needed to run the FM Radio app.

To hear the radio, choose the FM Radio app from the All Apps screen.

The first time you use the FM Radio app it scans for available stations, making a note of which are active. After the initial scan, you can use the FM Radio app to listen to broadcast FM radio on your phone. The app's interface is shown in Figure 16-6. Use the controls as illustrated in the figure to change stations or scan for new stations.

Exit FM Music app.

FM Radio notification

Headset connected

Current frequency

Scan

Station list

Skip to preset station.

Add station. | Scan

Skip to preset station.

Tune

Figure 16-6: FM Radio on the ThunderBolt.

You can listen to the FM Radio app over the ThunderBolt's speakers: Press the Menu soft button and choose the Speaker command.

To quit the FM Radio app, touch the Exit button, illustrated in Figure 16-6.

✔ You can do other things while you listen to FM Radio on your ThunderBolt.

✔ To return to the FM Radio app, pull down the notifications and choose FM Radio.

✔ The FM Radio app uses the headset as an antenna. If you unplug the headset, the FM Radio gets annoyed and asks you to reinsert the headset.

Listening to Internet radio

Though they're not broadcast radio stations, some sources on the Internet — *Internet radio* — play music. You can listen to this Internet music using the Slacker Personal Radio app that comes with your ThunderBolt.

Start Slacker from the All Apps screen. You need to create an account, if you don't already have one. Otherwise, log in to your account and then peruse the various stations available. From that point on, Slacker works just like listening to a portable radio.

Beyond Slacker, you can get other apps available for your ThunderBolt for listening to music as though the phone were a radio:

- Pandora Radio
- StreamFurious

Pandora Radio lets you select music based on your mood and customizes, according to your feedback, the tunes you listen to. The app works like the Internet site `www.pandora.com`, in case you're familiar with it.

StreamFurious streams music from various radio stations on the Internet. Though not as customizable as Pandora, StreamFurious uses less bandwidth.

Both apps are available at the Android Market. They're free, though a paid, pro version of StreamFurious exists.

- Listen to "Internet radio" when your phone is connected to the Internet via a Wi-Fi connection. Streaming music can use a lot of your cellular data plan's data allotment.

- See Chapter 18 for more information about the Android Market.

- Internet music of the type delivered by Slacker Personal Radio is referred to by the nerds as *streaming music.* That's because the music arrives on your ThunderBolt as a continuous download from the source. Unlike music you download and save, streaming music is played as it comes in and not stored long-term.

17

Other Useful Apps

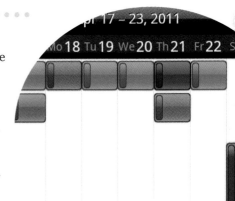

The ThunderBolt has unlimited potential, and the ability to replace just about every other gizmo you tote around. Think of any portable device that you'd regularly carry (except weaponry) and you probably have an app on the phone that accomplishes the same task. Most of these apps are probably included on the phone already, including the various and sundry apps discussed in this chapter. They all work to help you and your phone do amazing things.

Your Calculator

There's no need to carry a pocket calculator when you have a ThunderBolt, and no need to do any more math in your head. Simply saunter over to the All Apps screen and open the Calculator app.

In vertical orientation, the Calculator app looks like the traditional, handheld calculator. It has the standard buttons for division, multiplication, subtraction, and division, along with a C (Clear) button and a Backspace button.

In horizontal orientation, the Calculator app goes mathematically bonkers and adds scientific functions. Do not let them frighten you: Gently reorient the phone to an up-and-down position and the scary buttons go away.

- ✒ Plenty of alternative calculators are available on the Android Market. See Chapter 18 for directions on how to find them.

- ✒ One of my favorite Android calculators is Easy Calculator. I like it because it features parentheses, which make typing complex calculations a heck of a lot easier. It also sports a cheerful color scheme, which begs the question of why calculators feature such bleak designs.

- ✒ One common reason for using a calculator app is to calculate the tip that's due when dining out. Multiple free tip calculators are available at the Android Market. I have no particular one that I like more than others; simply search for *tip calculator* at the market and you see a bunch of them.

- ✒ To calculate the tip, enter the amount of the meal (not including taxes) and multiply by the percent as a decimal value. For example, for an 18 percent tip, use

```
amount × .18 =
```

Your Calendar

Make sure that you respond politely the next time someone gives you a date-book as a gift. It's the thought that counts, but, honestly, the ThunderBolt is a far more effective datebook, appointment calendar, and reminder utility than you'll find anywhere. It's all thanks to the Calendar app.

The Calendar app works in cahoots with the Google Calendar on the Internet. That way, your schedule is easily accessed from any computer or mobile gizmo that has access to the Internet. It makes you *want* to keep a schedule, if you know how the thing works.

- ✒ Google Calendar works with your Google account to keep track of your schedule and appointments. You can visit Google Calendar on the web at

```
http://calendar.google.com
```

- ✒ You automatically have a Google Calendar; it comes with your Google account.

TIP

✔ I recommend that you use the Calendar app on your ThunderBolt to access Google Calendar. It's a better way to access your schedule than using the Internet app to get to Google Calendar on the web.

✔ Before you throw away your datebook, copy into the Calendar app some future appointments and info, such as birthday and anniversaries.

Reviewing your schedule

To see what's happening next, to peruse upcoming important events, or just to know which day of the month it is, summon the Calendar app. It's found on the All Apps screen along with all the other apps that dwell on your ThunderBolt.

The Calendar app looks like a typical calendar, with the month and year at the top. Scheduled appointments and events appear on various days, as shown in Figure 17-1.

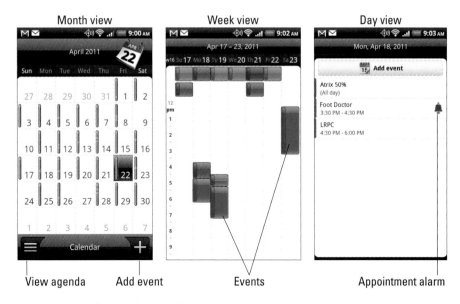

Figure 17-1: The Calendar's Month, Week, and Day views.

To switch between Month, Week, and Day views, press the Menu soft button and choose Month, Week, or Day. You can also choose the Agenda command to see all future upcoming events and appointments in list format.

To view your appointments by week, touch the Week tab. Or, you can touch the Day command to see your daily schedule. The List tab shows you all your events, one after the other, in a long list.

- ✒ See the later section "Making a new event" for information on reviewing and creating events.

- ✒ Use Month view to see an overview of what's going on, and use either Week or Day view to see your appointments as they happen during the day.

- ✒ I check Week view at the start of the week to remind me of what's coming up.

- ✒ Different colors flag your events to represent different calendars, or event categories, to which the events are assigned. See the later section "Making a new event" for information on calendars.

- ✒ Use your finger to flick the calendar displays left or right to see the next or previous day, week, or month.

Checking appointments and events

To see more detail about an event, touch it. When you're using Month view, touch the date with the event on it and then choose the event from the list.

The amount of detail you see depends on how much information was recorded when the event was created. Some events have only a minimum of information; others may have details, such as a location for the event. When the event's location is listed, you can touch that location and the Maps app pops up to show you where the event is being held.

Birthdays and a few other events on the calendar may be pulled in from your social networking sites. That probably explains why some events are listed twice; they're pulled in from two sources.

Making a new event

The key to making the calendar work is to add events: appointments, things to do, meetings, or full-day events such as birthdays or colonoscopies. To create a new event, follow these steps in the Calendar app:

1. **Select the day for the event.**

   Or if you like, you can switch to Day view, where you can touch the starting time for the new event.

2. **Press the Menu soft button and choose New Event.**

   If you're using Month view, you can touch the Add Event button, shown in Figure 17-1. For Week view, you can touch the approximate time on the day of the event, which saves you some time later.

The New Event screen appears. Your job now is to fill in the blanks to create the new event.

The more information you supply, the more detailed the event and the more you can do with it on your ThunderBolt as well as on Google Calendar on the Internet.

3. **Choose the Calendar event.**

Calendars help you keep your events organized, mostly because they're color-coded. The main calendar is the same as your Google account name. Other calendars can be created, though you can do that only on the Google Calendar website on the Internet.

4. **Type an event name.**

Sometimes I simply write the name of the person I'm meeting.

5. **Set the meeting duration using the From and To buttons.**

When the event lasts all day, such as a birthday or your mother-in-law's visit that was supposed to last for an hour, touch the All Day check mark.

To span an event over a few days, touch the All Day check mark and then list the starting and ending days.

6. **Input the event's location.**

This step is optional, though adding an event location not only tells you where the event will be located but also hooks that information into the Maps app. My advice is to type information into the Event Location field just as though you're typing information to search for in the Maps app.

When the event is displayed, the location can be found on the map by touching the icon next to the location on the event's summary screen.

7. **Type a description of the event.**

I use this field to type additional information, flight confirmation numbers, things to do at a meeting, additional directions, and stuff like that.

8. **Set whether the event has a reminder.**

The Calendar app is configured to automatically set a reminder ten minutes before an event begins. If you prefer not to have a reminder, touch the menu button and choose None as the alarm.

9. **Ignore the Guests field.**

Google uses email addresses in the Guests field to send invites or reminders to others. It can be handy; I have my son put my email address for events he creates that involve me. It's handy. But otherwise, you can ignore this field.

10. **Choose the repetition for the event, if any.**

    Most items you schedule are one-time events, so you can leave the button alone. Otherwise, touch it to set the repetition for an event. The repetition options available depend on the day of the week and day of the month of the original event.

11. **Touch the Save button to create the new event.**

You can change an event at any time: Simply touch the event to bring up more information and then press the Menu soft button to choose the Edit Event command.

To remove an event, long-press its entry and choose the Delete Event command. Touch the OK button to confirm.

✔ Use the Repetition button to create repeating events, such as weekly or monthly meetings, anniversaries, and birthdays.

✔ Reminders can be set so that the phone alerts you before an event takes place. The alert can show up as a notification icon (shown in the margin), or it can be an audio alert or a vibrating alert. Pull down the notifications and choose the calendar alert. You can then peruse pending events.

✔ To deal with an event notification, pull down the notifications and choose the event.

✔ Events also appear on the lock screen so that you see them immediately upon waking the phone. The notification doesn't go away until you dismiss the event.

✔ You can also create events by using the Google Calendar on the Internet. Those events are instantly synced with the calendar on your ThunderBolt.

Your Clock

Your ThunderBolt keeps constant, accurate track of the time, which is displayed at the top of every Home screen as well as on the lock screen. The display is lovely and informative, but it doesn't tell you the time in Sydney and New York. It can't wake you up in the morning, nor can it be used to time an egg. No, for these chores, you need the Clock app.

You find the Clock app dwelling on the All Apps screen. It's really five apps in one, as illustrated in Figure 17-2.

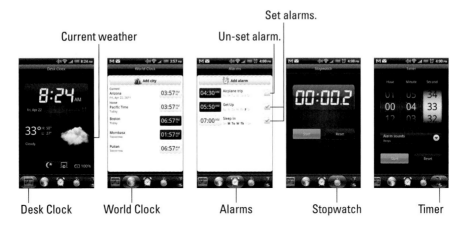

Figure 17-2: The Clock app and its Desk Clock and Alarms screens.

Alarm duties are accessed by touching the Alarms button, shown in Figure 17-2. To set a new alarm, follow these steps:

1. **After starting the Clock app, touch the Alarms button.**

2. **Touch the Add Alarm button.**

   The Set Alarm screen appears.

3. **Scroll the Hour, Minute, and AM/PM wheels to set the time for your alarm.**

4. **Type a description.**

   For example, type **Wake up for airport** or **End your nap now**.

5. **Touch the Alarm Sound item to choose which tone to hear when the alarm sounds.**

   If you're prompted, choose Android System to use the phone's sounds for the alarm. You can choose between preset alarm sounds or ringtones or pluck out a song from the Music library.

6. **Choose Repeat to set whether the alarm happens once, daily, or on specific days.**

7. **Keep the check mark by Vibrate.**

   Or, remove the check mark if you don't think a vibrating phone will rouse you.

8. **Touch the Done button to create and set the alarm.**

The alarm you create appears on the Clock app's Alarm screen.

Any new alarm you create is automatically set — it goes off when the proper time approaches. To disable an alarm, touch the green check mark to disable it. Disabling an alarm doesn't delete the alarm.

 The ThunderBolt alerts you to set alarms by displaying the Alarm Clock status icon, shown in the margin, in the phone's status area. The icon indicates that an alarm is set and pending.

Alarms must be set or else they don't trigger. To set an alarm, touch the gray square by the right of the alarm to place a green check mark there.

- ✔ To remove an alarm, press the Menu soft button and then choose which alarms you want to delete: Place a red X in the box by the alarm description. Touch the Delete button to remove the alarm.

- ✔ The alarm doesn't work when you turn off the ThunderBolt. The alarm does, however, go off when the phone is sleeping.

- ✔ A notification icon appears when an alarm has gone off but has been ignored.

Your eBook Reader

An *eBook* is an electronic version of a book. The words, formatting, figures, pictures — all that stuff is simply stored digitally so that you can read it on something called an eBook reader. For your ThunderBolt, the eBook reader is software that comes on the phone in the form of the Amazon Kindle app.

- ✔ The advantage of an eBook reader is that you can carry an entire library of books with you without developing back problems.

- ✔ Rather than buy a new book at the airport, consider getting an eBook instead, though you can still read a real book during takeoff and landing.

- ✔ Lots of eBooks are free, such as quite a few of the classics and including some that aren't that boring. Current and popular titles cost money, though the cost is often cheaper than the book's real world equivalent.

- ✔ Magazine and newspaper subscriptions are also available for eBook readers.

- ✔ You're not limited to using the Amazon Kindle app as your eBook reader. Other apps are available, including Aldiko, FBReader, Kobo, Laputa, and more. You can locate these eBook readers by perusing the Android Market, as described in Chapter 18.

Using the Amazon Kindle app

The good folks at Amazon recognize that you probably don't want to buy a Kindle eBook reader gizmo because you already have a nifty and portable do-everything device, the ThunderBolt. Therefore, the Amazon Kindle app serves as your eBook reader software and also provides access to the vast library of existing Kindle titles at Amazon.com.

Start the Amazon Kindle app by touching its icon, found on the All Apps screen.

The app is named *Amazon Kindle,* not just *Kindle.*

Upon starting the Amazon Kindle app, you see the registration screen. Log in using your email address and Amazon password.

Yes, you need an Amazon.com account to purchase eBooks (or even download freebies), so I highly recommend that you visit www.amazon.com to set up an account if you don't already have one.

After registering, or simply signing in, you see the main Kindle screen. If you're just starting out, your digital bookshelf is empty. See the next section for information on getting eBooks to read.

If you already have an Amazon Kindle account, after touching the Register button your ThunderBolt is synchronized with your existing library. To ensure that you get everything, press the Menu soft button and choose the Archived Items command.

Consider placing a shortcut to the Amazon Kindle app on the Home screen. See Chapter 22 for the complete, scintillating instructions.

Getting some reading material

An empty bookshelf is such a sad thing, especially when there are plenty of free books to be found. Whether you pay for a book or get a freebie, follow these steps in the Amazon Kindle app:

1. **Press the Menu soft button and choose Kindle Store.**

   You're transferred from the Amazon Kindle app into the Internet app. A special page opens for the Kindle store at Amazon.

2. **Search for the book you want, or browse the categories.**

   As a suggestion, choose the category Free Popular Classics to find something free to download.

3. **Touch to select a title.**

   Ensure that the title description says *Auto-delivered wirelessly.* That way, the title is automatically transferred (downloaded) to your phone.

4. **Touch the Buy Now with 1-Click button.**

5. **If you're getting a free title, or if you've already signed in to your Amazon account, skip to Step 10.**

6. **If prompted, sign in to your Amazon account.**

   Your email address is already filled in, but you need to touch the Password field and type your Amazon password.

7. **Touch the Sign In button.**

8. **If prompted to have the browser automatically remember your password, touch the Never button.**

   If you want, you can touch the Remember button to have the Internet app automatically recall your password. I recommend against it, for security reasons.

9. **If prompted, specifically when purchasing a paid eBook, choose the credit card you want to use for making the purchase.**

   For example, if you have multiple credit cards on file at Amazon, select the one to which you want to charge your eBook purchase.

10. **Touch the Read It Now button.**

    You return to the Amazon Kindle app on your phone.

The title you downloaded appears in the library shortly. If not, press the Menu soft button and choose the Sync command.

If you've already purchased eBooks from the Kindle store for another gizmo, you can archive them to your ThunderBolt: In the Amazon Kindle app, press the Menu soft button and choose the Archived Items command. You instantly see your existing Kindle library on the phone.

See the next section for information on reading eBooks.

 ✔ You receive an email confirmation message describing your purchase. The email appears even when you "buy" a free eBook.

 ✔ For a non-free eBook, you see a button titled Try a Sample. Touch that button to download a snippet from the title you're interested in. It's the equivalent of browsing at a bookstore (the real kind of bookstore).

 ✔ Not every title is available as an eBook.

Reading on your phone

After choosing a eBook in the Amazon Kindle app, you see it appear on the phone's touchscreen. For the first time you've opened a book, you see the first page. Otherwise, you're taken to the last page you were reading.

Figure 17-3 illustrates the basic reading maneuvers: You can touch the left or right sides of the screen to flip a page left or right, respectively. You can also swipe the pages left or right.

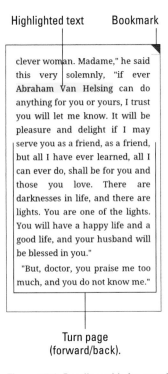

Figure 17-3: Reading with Amazon Kindle.

Touching a page in the upper right corner adds a bookmark, as shown in Figure 17-3. Touching the same spot again removes the bookmark.

When you touch the center of the page, you see an overlay that shows your location in the overall document. You can use the slider that appears at the bottom of the screen to quickly move to a new part of the document.

Long-press a word to activate highlighting mode, which works like selecting text on the ThunderBolt. (See Chapter 4.) Drag the starting and ending
markers to select text, and then choose whether you want to make a note,
highlight the text (refer to Figure 17-3), or search the text within the book or
at Wikipedia or on Google.

For more precise movement in an eBook, press the Menu soft button and
choose the Go To command. Use the Go To menu to choose where to go in
the document; if you choose the My Notes & Marks item, you can hop to one
of your bookmarks.

To return to the Kindle library (the main page), press the Menu soft button
and choose the Home command.

 ✔ You can change the size of the text on the page by pressing the Menu
   soft button and choosing the View Options menu. You choose from five
   preset text sizes.

 ✔ The book you're reading changes its orientation as you rotate the
   ThunderBolt, so turning the phone sideways may make the text easier
   for you to read.

Your Game Machine

One of the best ways to put expensive, high-tech gizmos to work is to play
games. Don't even sweat the thought that you have too much "business"
or "work" or other important stuff you can do on a ThunderBolt. The more
advanced the mind, the more the need for play, right? Indulge yourself.

Two sample games come with the ThunderBolt: Let's Golf 2, the golfing simu-
lation game shown in Figure 17-4, and Rock Band, which is an addicting, nostalgic music game.

Figure 17-4: A sample game on the ThunderBolt.

Both apps are teasers; neither is the full version of the game. You get just enough to whet your appetite. To get a full version, you have to pay for an upgrade, which each app happily and repeatedly reminds you to do.

Beyond the prepackaged apps is an entire universe of games, available at the Android Market.

See Chapter 18 for information on using the Android Market.

I have a few game suggestions on my website. Visit the Wambooli App-o-Rama:

www.wambooli.com/help/phone/app-o-rama

Your Flashlight

It's perhaps the silliest app on the ThunderBolt, but one you may find handiest. It's the Flashlight, and it allows you to manually control the phone's rear-camera flash LEDs so that the ThunderBolt can effectively light your way.

The Flashlight app makes sense. After all, I can't tell you how many times I've used my phone's touchscreen as a torch to illuminate my front door keyhole or a dark hallway or to forage for food when the refrigerator light is out.

Your Stock Ticker

Here's a stock tip for you: When everyone is crazy about stocks and the market is going up to the point that the laws of gravity no longer exist, it's time to get out of the market. Before then, you can enjoy the Stocks app, found on the All Apps screen.

The Stocks app taps into the Yahoo! Finance website, and it offers ways to look up market information, track stocks, and basically fret over your finances wherever you take the ThunderBolt.

 Because the ThunderBolt is a Google phone, you can use the Google Finance app for monitoring your favorite stocks, the market, or stock news. I prefer Google Finance over the Stocks app because I already use Google Finance on the web to keep track of stocks. All those stocks I follow using Google Finance on the web are magically synced with the Google Finance app on my ThunderBolt.

You can obtain Google Finance from the Android Market. See Chapter 18.

It's Your Video Entertainment

It's not possible to watch "real" TV on the ThunderBolt, but a few apps come close. The YouTube app is handy for watching random, meaningless drivel, which I suppose makes it a lot like TV. And then there's BitBop and Blockbuster, which let you buy and rent real movies and TV shows.

Enjoying YouTube

The Internet phenomenon *YouTube* proves that real life is indeed too boring and random for television. Or is it the other way around? Regardless, you can view the latest videos on YouTube — or contribute your own — by using the YouTube app on your ThunderBolt.

When you open the YouTube app, you see videos you've subscribed to. To visit the main YouTube screen, shown in Figure 17-5, press the Menu soft button and choose the Browse command. Then choose the All category.

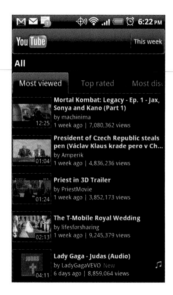

Figure 17-5: YouTube's All category.

To view a video, touch its name or icon in the list.

To search for a video, touch the Search soft button. Type or dictate what you want to search for and then peruse the results.

Videos in the YouTube app play at full-screen size in Landscape mode; otherwise, they play on the top part of the screen. Touch the screen to pause the video; touch again to play.

Press the Back soft button to return to the main YouTube app after watching a video or if you tire of a video and need to return to the main screen out of boredom.

✔ Use the YouTube app to view YouTube videos rather than use the Internet app to visit the YouTube website.

✔ Not all YouTube videos are available for viewing on mobile devices.

✔ Visit your account on YouTube by pressing the Menu soft button and choosing the My Channel command.

✔ When viewing your account (or channel), you can upload videos by touching the Upload button. Also see Chapter 14 for information on recording video on the ThunderBolt. Chapter 15 covers the specific steps for uploading (sharing) a video.

Buying and renting TV shows and movies

Two apps that come on the ThunderBolt allow you to buy or rent films and television shows. They're Bitbop and Blockbuster.

Bitbop is a subscription service available from Verizon's V CAST. It lets you watch popular TV shows without commercials any time you want, as long as you pay the subscription fee. As this book goes to press, it's $9.99 per month.

The Blockbuster app lets you rent or purchase mainstream movies to view on the ThunderBolt. The key is to have a Blockbuster account. When you do, you can follow the directions on the screen after starting the app and then get everything signed up and configured.

Your Weatherman

The Weather app, found near the end of the All Apps screen, is your daily link to weather information about your location and your favorite places in the world — well, favorite places that the Weather app recognizes. Like other Android weather apps, the one that ships with the ThunderBolt is rather limited to where it can prognosticate.

A better weather program is The Weather Channel. You can obtain at the Android Market by scanning the QR code in the margin or by visiting the Android Market directly, as described in Chapter 18.

⊯ The News app, found on the All Apps screen, doesn't give you daily headlines. Instead, it's a subscription reader for various websites. See Chapter 11 for details.

⊯ You can get various news apps for the ThunderBolt. Visit the Android Market to search for *news* and you'll find a slew of them. Be cautious: Some apps require a subscription or make you volunteer personal information before letting you read the news.

 ⊯ A useful app for sports fans is SportsTap. Use the bar code in the margin to get a copy.

18

Apps, Apps, and More Apps

Your ThunderBolt isn't limited to the paltry assortment of apps that were preinstalled by the evil phone company or the not-quite-as-evil-but-still-questionable manufacturer. No, you're free to add even more apps. The place to get those apps is the virtual Android Market shopping mall. Finding, installing, and maintaining those apps is the topic of this chapter.

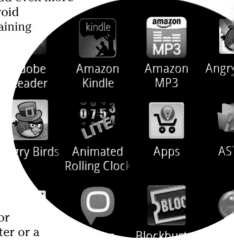

✔ Because the ThunderBolt uses the Android operating system, it can run nearly all the apps available for that operating system. That's over 100,000 apps.

✔ The only apps that show up in the Android Market are the apps that can run on your phone. Rest assured that you cannot obtain an app for your phone that is somehow incompatible.

✔ App is short for *application*. It's another word for software, or the programs that run on a computer or a mobile device, such as the ThunderBolt phone.

More Apps for Your Phone

The craving for new apps seems endless. Whether you get a recommendation from someone else, desire a game, or just want to see whether a specific type of app exists, the place you need to go is the Android Market. There you find the full variety of apps for your phone. You can search. You can browse. You can be amazed that your phone is capable of so much.

✔ The apps you obtain from the Android Market are *downloaded* into your phone. This transfer takes place over the digital cellular network or a Wi-Fi network.

✔ As far as speed goes, using the 4G network is the fastest way to download new apps from the Market into your phone. Even so:

✔ I highly recommend that you connect to a Wi-Fi network when you obtain apps at the Android Market. Using Wi-Fi doesn't incur data usage charges on your cellular plan. See Chapter 19 for details on connecting the ThunderBolt to a Wi-Fi network.

Going to the Market

You access the Android Market by opening the Market app, which can be found on the main Home screen as well as on the All Apps screen. After opening the Market app, you see its cheerful visage, similar to what's shown in Figure 18-1. You can browse for apps or games or for special apps from your cellular provider, by touching the appropriate doodad, as shown in the figure.

Find apps by browsing the lists: Choose Apps (refer to Figure 18-1). Then choose a specific category to browse. You can view apps by their price (paid or free) or find those that are "just in."

When you know an app's name or an app's category or even what the app does, searching for the app works fastest: Touch the Search button at the top of the Market screen (refer to Figure 18-1). Type all or part of the app's name or perhaps a description. Touch the Search button (the Magnifying Glass icon, to the right of the Search Android Market text box) to begin your search.

To see more information about an app, touch it. Touching the app doesn't buy it, but instead displays a more detailed description, screen shots, and comments, plus links to see additional apps or contact the developer.

Browse by category.

Browse all apps.   Search

Featured apps

Figure 18-1: The Android Market.

Return to the main Android Market screen at any time by pressing the Menu soft button and choosing the Home command.

✔ The first time you enter the Android Market, you have to accept the terms of service: Touch the Accept button.

✔ Pay attention to an app's ratings. Ratings are added by people who use the apps, like you and me. Having more stars is better. You can see additional information, including individual user reviews, by choosing the app.

✔ In addition to getting apps, you can download widgets for the Home screen as well as wallpapers for the ThunderBolt. Just search the Android Market for *widget* or *live wallpaper*.

✔ See Chapter 22 for more information on widgets and live wallpapers.

Getting a free app

After you locate an app you want, the next step is to download it. Follow these steps:

1. **If possible, activate the Wi-Fi connection.**

   Though the 4G LTE signal is just as fast (or faster) than Wi-Fi, this connection doesn't incur data usage charges. If you can get only a 3G connection, Wi-Fi is often faster.

   See Chapter 19 for information on connecting to a Wi-Fi network.

2. **Open the Market app.**

3. **Locate the app you want and open its description.**

   Refer to the preceding section for details. Notice that many apps are free.

4. **Touch the Free button.**

   The Free button is at the top of the screen, beneath the word *Install.*

5. **Touch the OK button to accept permissions and begin the download.**

   You return to the main Market screen as the app downloads. It continues to download while you do other things on your ThunderBolt.

    After the download is successful, the status bar shows a new icon, as shown in the margin. It's the Successful Install notification.

6. **Pull down the notifications.**

   See Chapter 3 for details, in case you've never pulled down notifications.

7. **Choose the app from the list of notifications.**

   The app is listed by its app name, with the text `Successfully Installed` beneath it.

At this point, what happens next depends on the app you've downloaded. For example, you may have to agree to a license agreement. If so, touch the I Agree button. Additional setup may involve signing in to an account or creating a profile, for example.

After the initial setup is complete, or if no setup is necessary, you can start using the app.

- The new app's icon can be found on the All Apps screen, along with all the other apps on your ThunderBolt.

- Newly installed apps are placed alphabetically by their names on the All Apps screen. Yes, that jiggles everything around.

✔ You can also place a shortcut icon for the app on the Home screen. See Chapter 22.

✔ The Android Market has many wonderful apps you can download. Chapter 26 lists some that I recommend, all of which are free.

Buying an app

Some great free apps are available, but many of the apps you dearly want probably cost money. It's not a lot of money, especially compared to the price of computer software. In fact, it seems odd to sit and stew over whether paying 99 cents for a game is "worth it."

I recommend that you download a free app first, to familiarize yourself with the process.

When you're ready to pay for an app, follow these steps:

1. **If possible, activate the ThunderBolt's Wi-Fi connection.**

2. **Open the Market app.**

3. **Browse or search for the app you want, and choose the app to display its description.**

   Review the app's price. It's priced in dollars, or whatever the local currency happens to be. Apps sold in a foreign currency show an approximate cost, as illustrated in Figure 18-1.

4. **Touch the button that lists the app's cost.**

   For example, the button may show US$0.99 as the cost. The button is found beneath the word *Buy*.

5. **Touch OK.**

   If you don't have a Google Checkout account, you're prompted to set one up. Follow the directions on the screen.

6. **Choose the payment method.**

   You can choose to use an existing credit card, add a new card, or — most conveniently — add the purchase to your cellular bill.

   If you choose to add a new card, you're required to fill in all information about the card, including the billing address.

7. **Touch the Buy Now button.**

   The Buy Now button has the app's price listed.

   After you touch the Buy button, the app is downloaded. You can wait or do something else with the phone while the app is downloading.

What about viruses?

Yes, it's true: Evil apps can be found in the Android Market. They're rare. So far, only a few dozen apps have been identified by Google as being malicious, and most of them were distributed overseas. These few apps in more than 100,000 is a small fraction, yet still the thought of obtaining a virus — or, more specifically, malware — is a scary one.

Malware, or *malicious* soft*ware,* is any program that either does something harmful or has the potential to do so. For your phone, that includes hijacking web pages, stealing personal information, raking up user fees (mostly from text messaging), or doing anything else that's unwanted, unintended, or undesired.

Generally speaking, check any app you install to ensure that it uses services that make sense.

You find those services listed on the screen where you touch the OK button to start the app's download. Make a note of any services the app uses that seem out of place. For example, a game that doesn't need to use the phone or text-messaging services but uses them anyway should raise an eyebrow. Peruse the reader comments. And use your brain: Avoid porn or "hacker" apps that claim to do things your phone shouldn't do, or which would otherwise cost you money.

The good news is that Google is aware of malware and helps you fight it. If your phone is infected and Google is aware of the problem, you may see a free update for the app named Android Market Security (or something similar). Install the app to clean up any infections.

The app may require additional setup steps, confirmation information, or other options.

 Successful installation of the app is flagged by the notification icon shown in the margin. You can pull down notifications to instantly launch your app. Otherwise, the app can be found on the All Apps screen, just like all other apps on your ThunderBolt.

Eventually, you receive an email message from Google Checkout, confirming your purchase. The message explains how you can get a refund from your purchase within 24 hours. The section "Removing an app," later in this chapter, discusses how it's done.

Basic App Management

There's more to your ThunderBolt apps than just running them. Beyond getting and installing new apps, you're occasionally asked to install an update. Or, maybe you want to remove an app that annoys you. If so, keep reading.

Reviewing installed apps

To peruse the apps you've installed on your ThunderBolt, follow these steps:

1. **Start the Market app.**

menu

2. **Press the Menu soft button.**

3. **Choose My Apps.**

4. **Scroll your downloaded apps.**

The list of downloaded apps should look similar to the ones shown in Figure 18-2. Instantly, you can see any apps that are in need of an update, as shown in the figure.

App needs updating.

Touch to see more info.

Install all nonmanual updates.

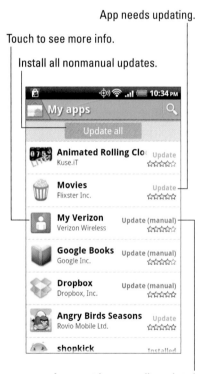

App must be manually updated.

Figure 18-2: Downloaded apps.

Touch an app to see the app's information screen (see Figure 18-3, a little later in this chapter). On the screen, you can read more about the app, open the app, uninstall the app, or configure automatic updating options. Later sections in this chapter describe the details.

- ✔ The list of installed apps shows all the apps you've downloaded on your ThunderBolt.

- ✔ Some apps on the All Apps screen might not be installed on your ThunderBolt. For example, they could have been downloaded and then removed. They remain in the list because you did, at one time, download the app.

- ✔ You may also see apps in the list that you have purchased or downloaded for other Android devices you might own. Look for those apps at the bottom of the list.

- ✔ To install on your ThunderBolt an app you've previously purchased or downloaded for another Android device, choose the app from the My Apps list and touch the Purchase or Install button.

Updating your apps

 One nice thing about using the Android Market to get your apps is that the Market also notifies you of new versions of the programs that are available: You see the Updates Available notification icon, shown in the margin.

Locate apps that need updating by pulling down the notifications and choosing Updates Available. Or, you can visit the My Apps list, as described in the preceding section.

When several apps require updating, and they've been configured for automatic updates, you see the Update All button (refer to Figure 18-2). Touch that button and all apps in the list that need updating are updated, one after the other. (Apps that require manual updating need to be updated one at a time.)

To update individual apps, or apps that require manual updating, follow these steps:

1. **Open the Market app.**

2. **Press the Menu soft button and choose My Apps.**

3. **Choose the app that needs to be updated.**

   In Figure 18-2, the My Verizon app requires a manual update, but the Animated Rolling Clock, Movies, and Angry Birds Seasons apps can be updated all at once:

4a. *When multiple apps need updating,* **touch the Update All button.**

4b. ***When only one app needs updating, or for a manual update,*** **choose the app and then touch the Update button; touch the OK button that replaces the Update button.**

Also see the next section on allowing apps to automatically update.

5. Touch the OK button in the warning dialog box to continue.

The app update is downloaded.

As when you initially install the app, you're free to do other things while your ThunderBolt apps are being updated.

 When downloading is complete, the Successful Install notification appears, as shown in the margin. When updating multiple apps, you see multiple Successful Install notification icons. Choose each notification to run the updated app, or touch the Clear button on the notifications panel to dismiss all Successful Install notifications.

 ✔ Just as when you first installed some apps, you may be prompted to agree (again) to the app's terms of services or licensing agreement, or you may be required to sign in to your account.

 ✔ When updating an app, the app is completely replaced by a new version. The original settings are retained, but an entirely new app is downloaded and installed.

 ✔ The Android operating system itself gets updated every so often. When an update is available, you see a message appear. You can choose to install the update now or later. I recommend installing all Android operating system updates as soon as you receive the notification.

Configuring automatic updating

Most of the time, you'll probably accept all updates offered by the Android Market. Rather than review the notifications every time, you can configure an app to be updated automatically. Here's how:

1. Start the Market app.

menu
2. Press the Menu soft button and choose the My Apps command.

3. Touch the app you want to configure.

The app's information screen appears.

4. Touch the box by Allow Automatic Updating.

The box shows a green check mark when automatic updating is enabled, as shown in Figure 18-3.

5. Press the Back soft button to return to the My Apps list, where you can configure automatic updating for other apps.

Automatic updating is on.

Remove the app.

Run the app.

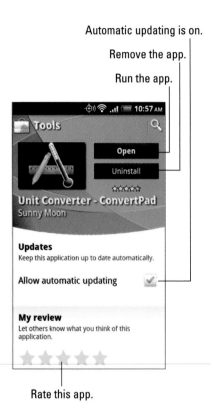

Rate this app.

Figure 18-3: An app's info screen.

Automatic updating isn't really automatic; you still have to touch the Update All button to update the apps. So, no app is ever automatically updated without your permission.

Some apps can't be updated automatically. It says so right beneath the Allow Automatic Updating prompt. Those apps must always be updated individually.

Uninstalling an update

When an update doesn't meet with your liking, you can roll it back. This situation might happen when you notice that the new version of an app doesn't seem to work right or you dislike the changes or improvements that come with the new version.

To uninstall an update, display the app's info screen (refer to Figure 18-3) and look for the Uninstall Updates button near the top part of the screen. Touch the Uninstall Updates button to remove the app's most recent update.

✔ Removing an app's updates sometimes removes the app itself.

✔ Not every app has the Uninstall Updates button available.

Removing an app

I can think of a few reasons to remove an app. It's with eager relish that I remove apps that don't work or somehow annoy me. It's also perfectly okay to remove redundant apps, such as when you may have multiple apps that do the same thing and you really use only one. Finally, removing apps frees up a modicum of storage in the ThunderBolt's internal storage area or the MicroSD card.

Whatever the reason, remove an app by following these directions:

1. **Start the Market app.**

2. **Press the Menu soft button and choose the My Apps command.**

3. **Touch the app that offends you.**

4. **Touch the Uninstall button.**

5. **Touch the OK button to confirm.**

   The app is uninstalled.

6. **Fill in the survey to specify why you removed the app.**

   The app is gone!

Even though the app is deleted (trust me), you can still find it near the bottom of the My Apps list. That's because you downloaded it once and the ThunderBolt remembers all apps you've ever downloaded.

✔ In most cases, if you uninstall a paid app before 24 hours has elapsed, your credit card or account is fully refunded.

✔ You can always reinstall paid apps that you've uninstalled. You aren't charged twice for doing so.

✔ You cannot remove apps that are preinstalled on the phone, by either HTC or Verizon. I'm sure there's probably a nerdy, technical way to uninstall the apps, but seriously: Just don't use the apps if you want to remove them and discover that you can't.

Advanced App Management

There are some areas beyond where the mere mortal ThunderBolt user would dare tread. Trust me: You can safely use your ThunderBolt, and install and manage your apps, without having to read this section. Still, you may be one of those few readers who finds this advanced app management information useful.

Getting app info

Beyond the My Apps list in the Market app lies a more technical place you can visit to monitor all apps on the ThunderBolt, as well as control a few more esoteric aspects of your apps. It's the Manage Applications screen, shown in Figure 18-4.

Apps you've installed

Apps that are currently running

All installed apps

Apps installed on MicroSD card

Figure 18-4: The Manage Applications screen.

To get to the Manage Applications screen, heed these directions:

1. **At the Home screen, touch the Menu soft button.**

2. **Choose Settings.**

3. **Choose Applications.**

4. **Choose Manage Applications.**

   You see a screen similar to the one shown in Figure 18-4.

5. **Touch an application to see its Application Info screen.**

   The Application Info screen is shown in Figure 18-5.

6. **Press the Back soft button to return to the Manage Applications screen.**

   Or, you can press the Home soft button to go directly to the Home screen.

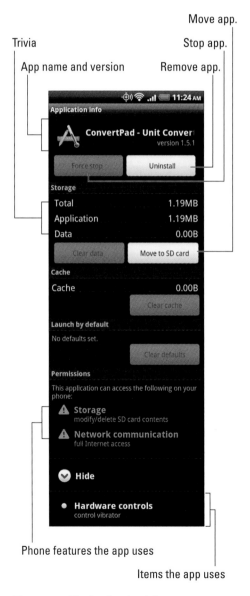

Move app.

Trivia

Stop app.

App name and version

Remove app.

Phone features the app uses

Items the app uses

Figure 18-5: The Application Info screen.

Touch the Uninstall button on the Application Info screen to uninstall an app, similar to the steps described in the section "Removing an app," earlier in this chapter. This technique works for those rare Android apps that don't appear on the My Apps list.

The Force Stop button is used to halt a program that runs amok. For example, I had to stop an older Android app that continually made noise and offered no option to exit. It was a relieving experience. See Chapter 23 for more details on shutting down apps run amok.

✔ Refer to the information in the Storage section (see Figure 18-5) to determine how much space the app is using in the ThunderBolt's internal storage area or MicroSD card.

✔ When an app is consuming a huge amount of space compared with other apps and you seldom use the app, consider it a candidate for deletion.

Moving an app

One of the more popular things you can do at the Application Info screen is move an app from the ThunderBolt's internal storage area to the MicroSD card: Touch the Move to SD Card button, shown in Figure 18-5, and the app is moved.

Conversely, to move an app from the MicroSD card to the phone's internal storage, touch the Move to Phone button, which replaces the Move to SD Card button for those apps that are installed on the MicroSD card.

To see which apps, if any, are already installed on the MicroSD card, touch the On SD Card tab, found at the top of the Manage Applications screen (refer to Figure 18-4).

✔ Some apps automatically install themselves on the MicroSD card. I don't know why, but they do.

✔ Moving an app is a hot topic on the various Android support forums on the Internet. I feel that you should keep apps in the ThunderBolt's main storage area as much as possible. One reason:

✔ If an app is placed on the MicroSD card, mounting the MicroSD card to your computer's storage system removes that app's shortcut from the Home screen. Refer to Chapter 20 for information on mounting the MicroSD card on your computer's storage system. See Chapter 22 for information on Home screen shortcuts.

Customizing the All Apps screen

Does it annoy you that new apps installed on the All Apps screen are sorted alphabetically? That means each app you install or remove messes up the scheme of things. Rather than accept that annoyance, you can customize and control how the All Apps screen presents information. Obey these steps:

1. **Touch the All Apps button on the Home screen to summon forth the All Apps screen.**

2. **Press the Menu soft button.**

мєnu

3. **Set the All Apps screen layout.**

   You can change whether the apps are shown in a grid or list, by choosing either the Grid or List command. Figure 18-6 shows both layouts.

Grid layout   List layout

Figure 18-6: Different ways to view the All Apps screen.

4. **While you're still viewing the All Apps screen, press the Menu soft button again.**

5. **Choose the Sort command.**

   Normally, and to much chagrin, apps on the All Apps screen sort alphabetically. As I've written elsewhere in this chapter, this sorting method messes things up when you add or remove apps. But there are, in fact, three options for listing your apps:

Alphabetical (A to Z): Choose this option if you're used to the standard way the ThunderBolt lists its apps.

Date (Most Recent): Choose this option to ensure that the apps you just added always appear at the top of the list.

Date (Oldest): Choose this option if you like consistency; new apps are added to the end of the list and their order stays consistent.

6. **Press the Home soft button to exit the All Apps screen.**

Though the best option for keeping the apps order consistent is Date (Oldest), it's no guarantee that your apps will always be where you left them. Removing an app juggles the list. The list can also change if you switch sorting methods. So if you're a real stickler for keeping apps in their same place, just resign yourself to the fact that you never achieve contentment when using the All Apps screen.

Part V
Nitty Gritty Details

The 5th Wave   By Rich Tennant

PCS PHONES

"So, what kind of roaming capabilities does this thing have?"

In this part . . .

And now, the boring part. Well, maybe not. But some of the things you do with your phone have nothing to do with making phone calls, getting email, browsing the web, finding places on the map, shooting pictures, watching video, playing games, or engaging in any of the fine and wonderful activities people have come to expect from cell phones. I've put all that remaining stuff in this part of the book.

In this part, you can find the details of using your phone — the nuts and bolts. The chapters in this part discuss changing settings on the phone, personalizing, using the phone when you're out and about, plus maintaining and troubleshooting. It's all important stuff, but it somehow lacks the thrill and adventure of all the other parts of the book. Because of that, I've added a word search puzzle at the end of Chapter 20. Enjoy.

It's a Wireless Life

In This Chapter

▶ Using the cellular data network

▶ Accessing a Wi-Fi network

▶ Using the ThunderBolt as a mobile hotspot

▶ Tethering the ThunderBolt connection

▶ Borrowing another gizmo's Internet connection

▶ Setting up Bluetooth

▶ Connecting and pairing to Bluetooth gizmos

*H*ere's a hot engineering tip: The key to creating a mobile device is to remove its wires. That's why you don't see many portable telephone poles. Truly mobile devices may use wires occasionally. There's a wire to charge the battery and perhaps a wire for synchronizing files. Maybe a wire as a fashion statement. Beyond that: No wires.

To meet power requirements, the ThunderBolt has a battery. For communications, the ThunderBolt takes advantage of the latest wireless networking technology. It includes being able to use the digital cellular network, Wi-Fi, and Bluetooth. In fact, the ThunderBolt is so excited about being wireless that it can even share its networking powers with other gizmos that are not so wirelessly blessed. This chapter covers the whole wireless deal.

The Wireless Networking World

The goal of wireless networking on your ThunderBolt is simple: Get on the Internet. Your phone can do that in two ways. The first is over the cellular data network, which is all that 4G stuff. The second way your phone can get on the Internet is the same way a computer can — over a Wi-Fi connection.

Understanding the cellular data network

You pay your cellular provider a handsome fee every month. The fee comes in two parts: One part is the telephone service; the second part is the data service, which is how the ThunderBolt gets on the Internet over the cellular provider's data network.

Several types of cellular data networks are available, based on the speed at which information is transmitted:

4G: The fourth generation of wide-area data network is as much as ten times faster than the 3G network, and is the latest craze in cellular networking. The HTC ThunderBolt is one of the first smartphones available that can access the 4G signal. Alas, the 4G data network isn't available in every location. Yet.

3G: The third generation of wide-area data networks is several times faster than the previous generation of data networks. This type of wireless signal is the most popular in the United States.

1X: Because the first-generation data network is slow and old, people despise it, despite the trendy *X* in its name. Still, it's better than nothing.

Other types of data networks are available, though these three are flagged by the ThunderBolt on the status bar. When digital information is being transmitted, the arrows in the network icon on the status bar become animated, indicating that data is being sent or received or both.

Your ThunderBolt always uses the best network available. So, if a 4G network signal is available, the phone uses it for Internet communications. Otherwise, the phone falls back to the 3G signal and occasionally to the 1X signal or nothing when no network signal is available.

✔ Accessing the digital cellular network isn't free. Your ThunderBolt most likely has some form of subscription plan for a certain quantity of data. When you exceed the quantity, the costs can become prohibitive.

✔ See Chapter 21 for information on how to avoid cellular data overcharges when taking your ThunderBolt out and about.

Turning on Wi-Fi

Wi-Fi is the same wireless networking standard used by computers for communicating with each other and the Internet. Making Wi-Fi work on your ThunderBolt requires two steps. First, you must activate Wi-Fi, by turning on the phone's wireless radio. The second step is connecting to a specific wireless network.

Follow these carefully written directions to activate Wi-Fi networking on your ThunderBolt:

1. **While viewing the Home screen, press the Menu soft button.**
2. **Choose Settings.**
3. **Choose Wireless & Networks.**
4. **Touch the square by the Wi-Fi option to place a green check mark next to it.**

The Wi-Fi radio is on, but the ThunderBolt isn't yet connected to a Wireless network. That's the second step, covered in the next section.

From the and-now-he-tells-us department, you can quickly activate the phone's Wi-Fi radio by touching the Wi-Fi widget. The widget is preinstalled on one of the ThunderBolt's Home screens, as shown in Figure 19-1.

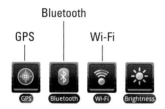

Figure 19-1: Turning on Wi-Fi with a widget.

To turn off Wi-Fi, repeat the steps in this section or touch the Wi-Fi widget, as shown in Figure 19-1. Turning off Wi-Fi disconnects the phone from any wireless networks.

- Using Wi-Fi to connect to the Internet doesn't incur data usage charges.
- The ThunderBolt Wi-Fi radio places an extra drain on the battery, but it's truly negligible. If you want to save a modicum of juice, especially if you're out and about and don't plan to be near a Wi-Fi access point for any length of time, turn off the Wi-Fi radio as described in this section.

Accessing a Wi-Fi network

After you activate the ThunderBolt's Wi-Fi radio, you can connect to an available wireless network. Heed these steps:

1. **Press the Menu soft button while viewing the Home screen.**
2. **Choose Settings.**
3. **Choose Wireless & Networks.**

4. **Choose Wi-Fi Settings.**

   The Wi-Fi screen appears.

5. **Ensure that Wi-Fi is on.**

   When you see a green check mark next to the Wi-Fi option, Wi-Fi is on. (This setting basically echoes the setting on the main Wireless & Networks screen.)

6. **Choose a wireless network from the list.**

   Available Wi-Fi networks appear at the bottom of the screen, as shown in Figure 19-2. When no wireless network is listed, you're sort of out of luck regarding wireless access from your current location.

   In Figure 19-2, I chose the Imperial Wambooli network, which is my office network.

Signal strength

Unprotected network

Password-protected network

Display notification when networks are nearby.

Wi-Fi radio is on.

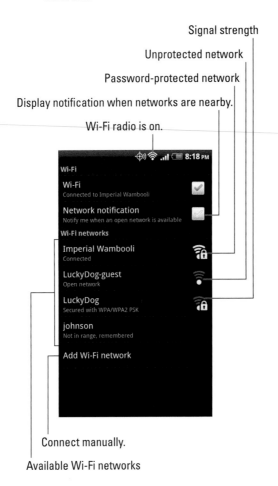

Connect manually.

Available Wi-Fi networks

Figure 19-2: Finding a wireless network.

7. **Optionally, type the network password.**

   If necessary, touch the Password text box to see the onscreen keyboard.

8. **Touch the Connect button.**

   You should be immediately connected to the network. If not, try the password again.

 When the ThunderBolt is connected to a wireless network, you see the Wi-Fi status icon, shown in the margin. This icon indicates that the phone's Wi-Fi is on and that the phone is connected and communicating with a Wi-Fi network.

Some wireless networks don't broadcast their names, which adds security but also makes connecting more difficult. In these cases, choose the command Add Wi-Fi Network (refer to Figure 19-2) to manually add the network. You need to input the network name, or *SSID,* and choose the type of security. You also need the password, if one is used. You can obtain this information from the girl with the pink hair who sold you coffee or from whoever is in charge of the wireless network at your location.

✔ Not every wireless network has a password.

✔ Some public networks are open to anyone, but you have to use the Internet app to get on the web to see the login page that lets you access the network: Simply browse to any page on the Internet and the login page shows up.

✔ If you place a check mark by the option Network Notification on the Wi-Fi Settings screen (refer to Figure 19-2), the phone alerts you to the presence of available Wi-Fi networks whenever you're in range and not connected to a network. This option is a good one to have set when you're frequently on the road. That's because:

✔ The ThunderBolt automatically remembers every Wi-Fi network it has ever been connected to and automatically reconnects upon finding the same network again.

✔ To disconnect from a Wi-Fi network, simply turn off Wi-Fi. See the preceding section.

 ✔ Unlike a cellular data network, a Wi-Fi network's broadcast signal goes only so far. My advice is to use Wi-Fi whenever you plan to remain in one location for a while. If you wander too far away, your phone loses the signal and is disconnected.

Share the Connection

You may grouse about your monthly cellular bill, but I'm sure you love the speed. Why not share that speed with other, less fortunate gizmos?

There are two tricks you can pull to share your ThunderBolt's überfast Internet connection with other devices. The first is to create a mobile hotspot, which allows any Wi-Fi–enabled gizmo to access the Internet by using your phone. The second is a direct connection between your phone and another device, which is a concept called *tethering*.

Creating a mobile hotspot

You can direct the ThunderBolt to share its digital cellular connection with as many as five other wireless gizmos. This process is referred to as *creating a mobile, wireless hotspot,* though no heat or fire is involved.

To set up a mobile hotspot on your ThunderBolt, heed these steps:

1. **Turn off the ThunderBolt's Wi-Fi radio.**

   You cannot be using a Wi-Fi connection when you create a Wi-Fi hotspot. Actually, the notion is kind of silly: If the ThunderBolt can get a Wi-Fi signal, other gizmos can too, so why bother creating a Wi-Fi hotspot in the first place?

2. **If you can, plug in the ThunderBolt.**

   It's okay if you don't find a power outlet, but running a mobile hotspot draws a lot of power. The phone's battery power drains quickly if you can't plug in.

3. **From the All Apps screen, open the Mobile Hotspot app.**

4. **Touch OK after reading (or not) the information screen.**

5. **Touch the box to place a green check mark by 4G Mobile Hotspot.**

   See Figure 19-3 for the location of the check mark.

6. **Touch the OK button in the Warning box.**

   You can, optionally, place a check mark by the option Do No Remind Me Again.

7. **Touch the button labeled Turn On Mobile Hotspot.**

   Directions on the screen explain how other people can connect their phones to your ThunderBolt's mobile hotspot. That's the information you use to connect to the mobile hotspot from another Wi-Fi gizmo — a computer or cell phone.

8. **Touch the OK button.**

9. **Touch the Home button to exit the Settings screen.**

   You can continue to use the ThunderBolt while it's sharing the digital cellular connection.

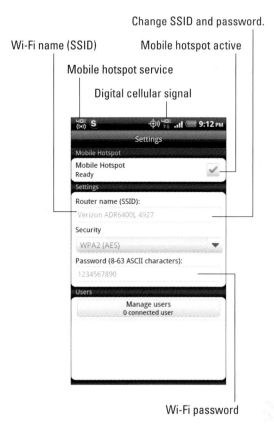

Figure 19-3: Configuring a mobile hotspot.

When the mobile hotspot is active, you see the Mobile Hotspot Service status icon appear, as shown in the margin. The icon reflects the service being shared, such as the 4G icon shown in the margin.

To turn off the mobile hotspot, pull down the Mobile Hotspot Service notification and remove the green check mark by 4G Mobile Hotspot (refer to Figure 19-3).

✔ Feel free to give your mobile hotspot a better name than the one that's preset, as shown in Figure 19-3. You can also assign a better password, while you're at it.

✔ The range for the mobile hotspot is about 30 feet.

✏ You may see extra charges for using the mobile hotspot. If your ThunderBolt has a measured data plan, be careful! Those per-megabyte fees can add up quickly when the mobile hotspot is active.

✏ Don't forget to turn off the mobile hotspot when you're done using it.

Sharing the Internet connection

You can share an Internet connection with your ThunderBolt phone in two ways. Both methods involve connecting the phone to another device, such as a computer, by using the USB cable. It's probably because of that cable that the term *tethering* is used.

The two types of shared Internet connection are

Internet Connection Mode: In this mode, the phone's Internet connection is shared with another device.

Internet Pass-Through: In this mode, the phone uses the other device's Internet connection.

So, basically, the first method shares another device's Internet connection with the phone, and the second method shares the phone's connection with the other device.

The easiest way to set up the connection is simply to plug the phone into the other device via a USB cable. When you do, you see the Connect to PC Screen, shown in Figure 19-4. You can immediately choose a sharing option, as shown in the figure.

The Connect to PC Screen goes away after a few moments. If you miss it, you can share the Internet connection by following these steps when the phone is connected by USB cable to another device:

1. **At the Home screen, press the Menu soft button.**

2. **Choose Settings.**

3. **Choose Wireless & Networks.**

4. **At the bottom of the screen, choose either Internet Connection Mode or Internet Pass-Through.**

The following subsections describe what happens next.

✏ The Connect to PC screen, shown in Figure 19-4, is also used to synchronize files between the ThunderBolt and a computer. See Chapter 20.

✔ Sharing the digital network connection incurs data usage charges against your cellular data plan. Be careful with your data usage when you're sharing a connection.

✔ You may be prompted on the PC to locate and install software for the ThunderBolt. Do so: Accept the installation of new software when prompted by Windows.

✔ I have not been successful at using either tethering method between the ThunderBolt and a Macintosh computer.

Change only (no Internet or data sync)

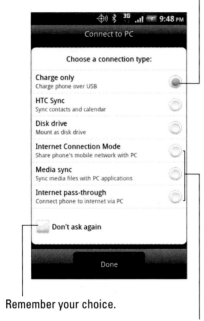

Remember your choice.

Internet sharing options

Figure 19-4: Choosing a connection type.

Internet Connection Mode

To share the ThunderBolt's digital cellular connection with a computer or another gizmo, follow Steps 1 through 4 earlier in this section and choose Internet Connection Mode.

The connection should be instantaneous on a Windows computer. You have to check the network connections in the Windows Control Panel to confirm that a new network is available and being used to access the Internet.

To end Internet Connection Mode, repeat Steps 1 through 4 and remove the green check mark by Internet Connection Mode.

- ✔ Ensure that the phone's Wi-Fi is turned off. You can share only the digital cellular connection, not the Wi-Fi connection, with another device.
- ✔ Oddly enough, no notification icon appears and lets you know that the other device is using your ThunderBolt's digital cellular connection.

Internet Pass-Through

To have the ThunderBolt use another device's Internet connection, connect the phone to the other device (assuming that the other device has an Internet connection) and follow Steps 1 through 4 earlier in this section. Choose Internet Pass-Through and then continue with these steps:

5. **Read the firewall warning.**

   If the other device is using a firewall, you need to allow access by the ThunderBolt.

6. **Touch the OK button to dismiss the Connection Troubleshooting warning.**

7. **If prompted, install the driver software.**

8. **Use the Internet connection.**

After your phone is connected, the Internet Pass-Through notification icon appears, as shown in the margin. The icon reminds you that the phone is using the other device's Internet connection.

To end Internet Pass-Through mode, pull down the notifications and choose Internet Pass-Through. Remove the green check mark.

If your computer can't find the proper drivers, download the HTC Sync program. Detailed directions can be found in Chapter 20, though, for now, here's the website address to type in your computer's web browser:

```
www.htc.com/www/support.aspx
```

Choose the link to download the HTC Sync application for all HTC Android phones.

The Bluetooth Thing

It has nothing to do with teeth and even less to do with the color blue. *Bluetooth* is a wireless way to connect various devices, which in the computer world would be called *peripherals.* For your ThunderBolt, the primary Bluetooth peripheral you'll consider using is one of those implant-like earphones. But you can do far more with Bluetooth than walk around and look like a Borg.

Understanding Bluetooth

Bluetooth was invented by nerds as a way to add a variety of peripherals to a computer without all the mess of wires. The entire Bluetooth gamut of gizmos includes mice, keyboards, monitors, speakers, printers, and other devices I can't think of right now.

Because Bluetooth is a wireless standard, a bit more work is involved in connecting a Bluetooth device than a wired device. For a wired device, you connect the wire. For a Bluetooth device, you go through these general gyrations:

1. **Activate the Bluetooth wireless radio on each gizmo.**

   There are two Bluetooth gizmos: the peripheral and the main device to which you're connecting the gizmo, such as the ThunderBolt.

2. **Make sure that the gizmo you're trying to connect is discoverable.**

   By making a Bluetooth device *discoverable,* you're telling it to say "Here I am" to other Bluetooth devices.

3. **On the main device, such as your phone, examine the list of available Bluetooth devices.**

4. **Choose the peripheral device you want to pair with your phone.**

   This action is known as *pairing* the devices.

5. **Optionally, confirm the connection on the peripheral device.**

   For example, you may be asked to input a code or press a button.

6. **Use the device.**

   What you can do with the device depends on what it does.

When you're done using the device, you simply turn it off. Because the Bluetooth gizmo is paired with the ThunderBolt, the gizmo is automatically reconnected the next time you turn it on (that is, if you have Bluetooth activated on the phone).

Bluetooth devices are marked with the Bluetooth icon, shown in the margin. It's your assurance that the peripheral can work with other Bluetooth devices.

 Bluetooth was developed as a wireless version of the old RS-232 standard, the serial port on early personal computers. Essentially, Bluetooth is wireless RS-232, and the variety of devices you can connect to, as well as the things you can do with Bluetooth, are similar to what you could do with the old serial port standard.

Turning on Bluetooth

To make the Bluetooth connection, you turn on the ThunderBolt's Bluetooth radio. Obey these directions:

1. **From the All Apps screen, choose the Settings icon.**

2. **Choose Wireless & Networks.**

3. **Place a green check mark by the Bluetooth option.**

The easier way, of course, is to use the Bluetooth widget, shown in Figure 19-1. Simply touch the Bluetooth button to activate Bluetooth on the ThunderBolt.

 When Bluetooth is on, the Bluetooth status icon appears, as shown in the margin.

To turn off Bluetooth, repeat the steps in this section: Remove the check mark or touch the Bluetooth widget.

Connecting to a Bluetooth peripheral

To make the Bluetooth connection between the ThunderBolt and another gizmo, follow these steps:

1. **Ensure that Bluetooth is on.**

   Refer to the preceding section.

2. **Turn on the Bluetooth gizmo and ensure that its Bluetooth radio is on.**

   Some Bluetooth devices have separate power and Bluetooth switches.

3. **On the ThunderBolt, press the Menu soft button from the Home screen and choose the Settings command.**

4. **Choose Wireless & Networks.**

5. **Choose Bluetooth Settings.**

6. **If the other device has an option to become visible, or discoverable, select it.**

   For example, some Bluetooth gizmos have a tiny button to press that makes the device visible to other Bluetooth gizmos. (You don't need to make the ThunderBolt visible unless you're accessing it from another Bluetooth gizmo.)

7. **Choose Scan for Devices.**

   Eventually, the device should appear on the Bluetooth Settings screen, in the Bluetooth Devices area, as shown in Figure 19-5.

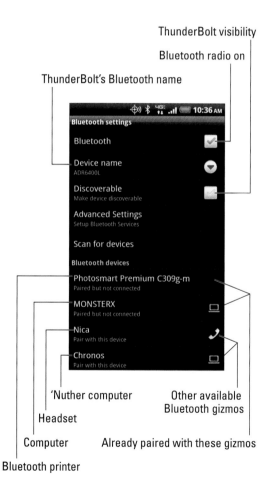

ThunderBolt visibility

Bluetooth radio on

ThunderBolt's Bluetooth name

'Nuther computer

Other available Bluetooth gizmos

Headset

Computer

Already paired with these gizmos

Bluetooth printer

Figure 19-5: Finding Bluetooth gizmos.

   8. **Choose the device.**

   9. **If necessary, input the device's passcode or otherwise acknowledge the connection.**

      Not every device had a passcode. If prompted, acknowledge the passcode on either the ThunderBolt or the other device.

   After you acknowledge the passcode (or not), the Bluetooth gizmo and your ThunderBolt are connected and communicating. You can begin using the device.

    The Bluetooth status icon changes when a Bluetooth device is paired and connected, as shown in the margin. It's your clue that the Bluetooth connection is active and working.

   To break the connection, you can either turn off the gizmo or disable the Bluetooth radio on your ThunderBolt. Because the devices are paired, when you turn on Bluetooth and reactivate the device, the connection is instantly reestablished.

   ✔ You can unpair a device by long-pressing the device on the Bluetooth Settings screen (refer to Figure 19-5) and choosing the Disconnect and Unpair command from the menu that pops up.

   ✔ There's normally no need to unpair a device unless you no longer plan on using it ever again.

   ✔ Bluetooth devices are disconnected by simply turning them off. They remain paired unless you specifically unpair them.

    ✔ When a Bluetooth headset, or ear gizmo, is connected to the ThunderBolt, you press its button to answer a call. While you're using a Bluetooth headset, the Phone Call notification shows up tinted blue, shown in the margin. Also, a special Bluetooth button appears on the call screen, which lets you quickly switch between the headset and the phone's speaker.

   ✔ Chapter 15 offers information on using Bluetooth to share files between your ThunderBolt and a Bluetooth-equipped computer. It also describes how you can print photographs on a Bluetooth printer.

    ✔ If you need help configuring Bluetooth for file exchange with a computer, touch the Advanced Settings command on the Bluetooth Setting screen (refer to Figure 19-5). Ensure that there's a green check mark in the box by FTP Server.

✔ I confess that using Bluetooth isn't the easiest thing. It requires patience to set things up and get the gizmos talking. Too often, you may find devices "paired but not connected." Some devices, such as printers and certain computers, may lack the drivers (control software) to complete the connection.

✔ Bluetooth can use a lot of power. Especially for battery-powered devices, don't forget to turn them off when you're no longer using them with the ThunderBolt.

20

Connect and Synchronize

In This Chapter

▷ Connecting the phone to your computer

▷ Setting USB connection options

▷ Disconnecting from a computer

▷ Synchronizing media with doubleTwist

▷ Using HTC Sync

▷ Swapping files manually

*G*oing against the grain of the whole wireless/portable notion is the ugly truth that the ThunderBolt does, from time to time, employ wires. And, if you're going to use a wire, it might as well be the USB cable that came with the phone. The USB cable can be used to charge the phone, but more importantly to share and synchronize files between your computer and the ThunderBolt. It's a necessary job, but it's not the easiest thing to do.

..connection type:

.e over USB ◯

.c
tacts and calendar ◯

ive
disk drive ◉

Connection Mode ◯
s mobile network with PC

aplications ◯

The USB Connection

The most direct way to connect a ThunderBolt to a computer is to use the USB cable that came with the phone. It's a match made in heaven, but like many matches, it often doesn't work smoothly. Rather than hire a counselor to put the phone and computer on speaking terms, I offer you some good USB-connection advice in this section.

Connecting the ThunderBolt to your computer

The USB connection between the ThunderBolt and your computer works faster when both devices are physically connected, and it's the USB cable that does the job. Like nearly every computer cable in the Third Dimension, the USB cable has two ends:

> ✓ The A end of the USB cable plugs into the computer.
>
> ✓ The other end of the cable plugs into the bottom of the ThunderBolt.

The connectors cannot be plugged in either backward or upside down.

After you understand how the cable works, plug the USB cable into one of the computer's USB ports. Then plug the USB cable into the ThunderBolt. See the next section for information on what happens next.

> ✓ When the ThunderBolt is connected via USB cable to a computer, you see the USB connection notification icon appear. Pull down the notifications to confirm or review the type of USB connection being used.
>
> ✓ Though the term *PC* is shown on the ThunderBolt screen, the other computer can be a Macintosh. Even so:

> ✓ The Macintosh operating system doesn't like the Verizon Mobile CD utility built into the ThunderBolt. Because the phone always tries to mount the pseudo-CD drive when it's connected to a computer (thank you, Verizon), you see an error message on your Macintosh when you first connect the ThunderBolt. Click the Eject button to dismiss the error message.
>
> ✓ You can still use the ThunderBolt's USB connection with a Macintosh. I recommend that you use the Disk Drive setting.
>
> ✓ You may see a lot of activity when you first connect the ThunderBolt to a Windows PC. Notifications might pop up about new software that's installed, or you may see the AutoPlay dialog box prompting you to install software. Do so.

Configuring the USB connection

Upon gleeful connection of your ThunderBolt to a computer, you see the Connect to PC screen, shown in Figure 20-1. You have three options on the screen for sharing and synchronizing files with the computer, as illustrated in the figure:

HTC Sync: Choose this option only when using the HTC Sync software on your Windows PC. See the section "Using HTC Sync," later in this chapter.

Disk Drive: Choose this option to add the ThunderBolt's MicroSD card to your computer's storage system, similar to plugging in a thumb drive or media card. Well, actually, it's *exactly* like plugging in a media card.

Media Sync: Choose this option to fool your computer into believing that the ThunderBolt is an MP3 music player or digital camera. It's the ideal option for swapping music or pictures between the phone and your computer because many music and photo sharing programs instantly recognize the phone when it's connected by using this option.

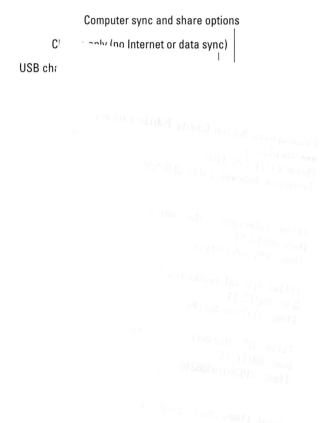

Computer sync and share options

C͟ ͟ ͟ ͟ ͟only (no Internet or data sync)

USB ch͟

sn't involve swapping files, one uses the computer's -don't-know-what-else-to-

creen are covered in

- You can change the connection type by pulling down the notifications and choosing the USB option from the top of the list: Change Connection Type. The Connect to PC screen appears (refer to Figure 20-1). Use it to choose a new connection option.

- No matter which USB connection option you choose, the ThunderBolt battery charges when it's connected to a computer's USB port — as long as the computer is turned on, of course.

- If you have a Macintosh, use the Disk Drive setting.

- When you're using the Disk Drive USB option, the ThunderBolt appears as a drive on a Windows computer, assigned a drive letter and available in the Computer window. Or, on a Macintosh, the ThunderBolt appears as a generic removable drive icon, named NO NAME, on the right side of the screen.

- When you choose the Media Sync type of USB connection, an animated Media Sync notification icon appears, similar to the one shown in the margin, though the icon on your ThunderBolt spins around.

- Refer to Chapter 16 for information on copying music from your computer to the ThunderBolt.

- You cannot access the phone's MicroSD card while the ThunderBolt is mounted into a computer storage system. Items such as your music and photos are unavailable until you unmount the MicroSD card from the computer.

Breaking the USB connection

You don't just yank out the USB cable when you're done using the ThunderBolt with your computer. That would be a Bad Thing. I'm serious: Unplugging a USB cable without properly unmounting the ThunderBolt can damage the phone's MicroSD card. Rather than foolishly risk such a thing, ensure that you properly dismount the phone from the computer by following these steps:

1. **Close whichever window or programs are accessing the ThunderBolt from the computer.**

2. **Properly unmount the phone from the computer's storage system.**

   On a PC, locate the phone's icon in the My Computer window. Right-click the icon and choose the Eject or Safely Remove command.

   On a Macintosh, drag the phone's disk drive icon to the Trash.

3. **On the ThunderBolt, pull down the notifications and choose the USB notification.**

4. **Select Charge Only.**

   The phone is now safely removed from the computer, though you can take it one step further, if you like:

5. **Disconnect the phone from the computer.**

   Unplug the USB cable.

After you unmount the ThunderBolt, its MicroSD card once again becomes available to the apps and you can freely browse your music and photos.

That Syncing Feeling

Most of the time, moving files between your phone and your computer is an automatic operation. Especially when it comes to media files, such as pictures, video, and music, the solution for copying files is something called *synchronization*. Using special software, you can quickly copy media between your phone and computer — if you know where to find the software and how to use it.

✔ Pictures can be shared between your ThunderBolt, computer, and the Internet when you use the Flickr website, covered in Chapter 15.

✔ Also see Chapter 16 for additional information on synchronizing music between your computer and phone.

Synchronizing media with doubleTwist

The doubleTwist utility is a popular way to synchronize music, photos, videos, and more between your computer and the ThunderBolt. After all, computers are all about making things easier, right?

The good news is that doubleTwist is free. It's not an Android app, but, rather, a program you download for your computer. Visit `www.double twist.com` to grab a copy for your PC or Macintosh.

The bad news is that doubleTwist is having issues with connecting a PC to a ThunderBolt as this book goes to press. That's okay: See the next section on using HTC Sync on a Windows computer.

After installing the doubleTwist program, run it. Connect the ThunderBolt to your computer using the USB Disk Drive option, discussed earlier in this chapter.

After everything is hunky and dory, you can peruse media on your computer and use the mouse to drag and drop items to the ThunderBolt. Likewise, you can drag and drop media from the ThunderBolt to your computer. See Figure 20-2.

The ThunderBolt

View local media.　　　　　　　　　View your iTunes music library.

Look for apps.　Copy a song by dragging it to the ThunderBolt.

Locked files cannot be copied.

Figure 20-2: The doubleTwist synchronization utility.

When you're done pilfering media from your computer, quit doubleTwist. Remember to properly unmount the ThunderBolt from the computer before you yank out the USB cable. On a Mac, drag the ThunderBolt's drive icon to the Trash to properly unmount it.

- ✔ You cannot copy media purchased at the iTunes store from the Mac to the ThunderBolt. Apparently, you need to upgrade to iTunes Plus before this operation is allowed.

- ✔ doubleTwist doesn't synchronize contact information.

- ✔ A doubleTwist app is available, but you don't need it in order to synchronize files with your computer. The doubleTwist app is a media player app.

- ✔ For a mere $4.99, you can purchase the doubleTwist Airsync app, which allows you to use the Wi-Fi network to synchronize files between your ThunderBolt and the computer. Of course, the USB connection is free.

- ✔ The error message generated on a Mac when you connect the ThunderBolt has to do with the silly Verizon Mobile CD drive.

Using HTC Sync

If you have a Windows PC, you can use the HTC Sync program to synchronize pretty much all the information stored on your ThunderBolt with your computer. It all starts by obtaining the HTC Sync program. Heed these steps:

1. **On your PC, visit the website www.htc.com/www.support.aspx.**

   Type the address into your PC's browser to visit the site.

2. **On the HTC Support website, click the link to install HTC Sync for HTC Android Phones & HTC Smart.**

   The remaining steps may change if HTC modifies its website after this book goes to press.

3. **Choose the HTC Sync application from the menu button near the bottom of the web page.**

4. **Click the Download button.**

5. **Run the program you downloaded to install the HTC Sync program on your PC.**

   Continue obeying the directions on the screen until the software is installed.

The HTC Sync program doesn't do anything until you connect your ThunderBolt to the computer. When you do, and the Connect to PC screen appears (refer to Figure 20-1), choose the option HTC Sync. At this point, the HTC Sync program on your computer comes to life and synchronization instantly takes place.

Figure 20-3 illustrates the HTC Sync interface. You must turn on individual categories of items to sync. You can use the big, green Sync Now button to manually sync files at any time.

When you're done synchronizing files, click the Disconnect button, shown in Figure 20-3. You can then unmount the ThunderBolt, as described in the section "Breaking the USB connection," earlier in this chapter.

 ✔ Your phone's contacts and calendars are automatically synchronized with Google on the Internet.

 ✔ Don't confuse HTC Sync with the *HMS Sink,* which was a ship that briefly served in His Majesty's Navy back in World War I.

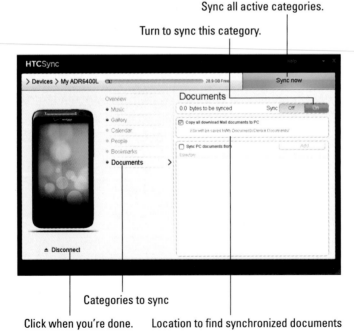

Sync all active categories.

Turn to sync this category.

Categories to sync

Click when you're done.   Location to find synchronized documents

Figure 20-3: HTC Sync in action.

Synchronizing manually

When you can't get software on your computer to synchronize automatically, you have to make the old manual connection. Yes, it can be complex. And bothersome. And boring. But it's often the only way to get some information out of the ThunderBolt and on to the computer, or vice versa.

Follow these steps to set up a file transfer from your computer to the ThunderBolt:

1. **Connect the ThunderBolt to the computer by using the USB cable.**

2. **Mount the ThunderBolt as a disk drive.**

3a. **On a PC, in the AutoPlay dialog box, choose the option Open Folder to View Files.**

   The option might also read Open Device to View Files.

   You see a folder window appear, which looks like any other folder in Windows. The difference is that the files and folders in that window are on the ThunderBolt, not on your computer.

3b. **On a Macintosh, open the Removable Drive icon that appears.**

   The ThunderBolt is assigned a generic, removable drive icon when it's mounted on a Macintosh. Most likely, it's given the name NO NAME.

4. **Open a folder window on your computer.**

   It's either the folder from which you're copying files to the ThunderBolt or the folder that will receive files from the ThunderBolt. For example, it might be the Documents folder.

   If you're copying files from the ThunderBolt to your computer, use the Pictures folder for pictures and videos and use the Documents folder for everything else.

5. **Drag the file icons from one folder window to the other to copy them between the phone and computer.**

   Use Figure 20-4 as your guide.

6. **When you're done, properly unmount the ThunderBolt from your computer's storage system and disconnect the USB cable.**

   You must eject the ThunderBolt's drive icon from the Macintosh computer before you can turn off USB storage on the phone.

Specific folders on the ThunderBolt

ThunderBolt is Drive F on this PC.

Drag files to here to copy to the root.

Figure 20-4: Copying files to the ThunderBolt.

Any files you've copied to the phone are now stored on the ThunderBolt's MicroSD card. What you do with them next depends on the reasons you copied the files: to view pictures, use the Gallery, import vCards, use the Contacts app, listen to music, or use the Music Player, for example.

- Files you've downloaded on the ThunderBolt are stored in the `Download` folder.

- Pictures and videos on the ThunderBolt are stored in the `DCIM/100Media` folder.

- Music on the ThunderBolt is stored in the `Music` folder, organized by artist.

- Quite a few files can be found in the *root folder,* the main folder on the ThunderBolt's MicroSD card, which you see when the phone is mounted into your computer's storage system and you open its folder.

ThunderBolt word search puzzle

T	E	N	O	Z	I	R	E	V	M	A	C
E	T	P	A	E	E	B	U	T	U	O	Y
S	A	H	Q	P	M	R	G	S	S	O	A
T	T	O	U	O	A	O	A	I	I	G	C
C	W	N	W	N	N	W	H	W	C	A	I
A	E	E	S	K	D	Z	Y	T	M	M	N
T	L	O	W	U	R	E	A	E	E	E	T
N	L	O	Y	S	O	R	R	L	L	S	E
O	I	A	P	B	I	A	A	B	G	G	R
C	A	L	E	N	D	A	R	U	O	X	N
Y	M	W	A	M	A	P	S	O	O	L	E
S	G	O	S	K	Y	P	E	D	G	A	T

ANDROID	CONTACTS	HOME	PHONE	VERIZON
APP	DOUBLETWIST	HTC	SKYPE	WEB
BROWSER	GAMES	INTERNET	SWYPE	YOUTUBE
CALENDAR	GMAIL	MAPS	THUNDERBOLT	
CAMERA	GOOGLE	MUSIC	USB	

✔ A good understanding of basic file operations is necessary to get the most benefit from transferring files between your computer and the ThunderBolt. These basic operations include copying, moving, renaming, and deleting. It also helps to be familiar with the concept of folders. A doctorate in entanglement theory is optional.

21

Have Phone, Will Travel

*A*s a mobile device, the ThunderBolt is designed to be with you wherever you go. You truly have no limit on where you can take the phone, though I'm not certain about outer space. Still, it's possible to wander all over Earth with your ThunderBolt. As long as the phone is charged and you can get a signal, it should work.

Then again, how the phone works when you wander far and wide is a good question. You should consider various issues regarding taking your ThunderBolt beyond the reach of your subscribed cellular network. You should also consider issues about taking your phone abroad. This chapter addresses those concerns — plus, it offers various tips and suggestions for making the ThunderBolt a seasoned traveler and international raconteur.

The Perils of Roaming

The whole concept of roaming seems rather droll. You may have thoughts of the song *Home on the Range,* where the buffalo roam and the deer and the antelope play. I doubt that those ungulates would be as eager if their roaming incurred cell phone surcharges.

On your ThunderBolt, *roaming* implies that your phone has wandered from your cellular provider's network and is accessing another cellular network. The other cellular network is more than happy to provide you with service — at a steep rate.

The ThunderBolt alerts you whenever you're roaming. You see the Roaming icon appear at the top of the screen, in the status area, similar to the one shown in the margin. The icon tells you that you're outside the regular signal area, possibly using another cellular provider's network.

When the phone is roaming, you may see the text *Emergency Calls Only* displayed on the locked screen.

Turning off roaming data services

To avoid roaming surcharges, you can simply turn off the phone, or decline all calls, while you're outside your cellular provider's service area. You can, however, altogether avoid using the other network's data services while roaming. Follow these steps:

1. **On the Home screen, press the Menu soft button and choose Settings.**

2. **Choose Wireless & Networks.**

3. **Choose Mobile Networks.**

4. **Ensure that the Data Roaming option isn't selected.**

   Remove the green check mark by the Data Roaming option.

The ThunderBolt can still access the Internet over the Wi-Fi connection when you're roaming. Setting up a Wi-Fi connection doesn't make you incur extra charges, unless you have to pay to get on the wireless network. See Chapter 19 for more information about Wi-Fi.

Disabling multimedia text messages when roaming

Another network service you might want to disable while roaming has to do with multimedia text messages, or *MMS*. To avoid surcharges from another cellular network for downloading an MMS message, follow these steps:

1. **Open the Messages app.**

2. **If the screen shows a specific conversation, press the Back soft button to return to the All Messages screen.**

   (It's the screen that lists all your conversations.)

3. **Touch the Menu soft button and choose Settings.**

4. Remove the green check mark by Roaming Auto-Retrieve.

Or, if the item isn't selected, you're good to go — literally.

For more information about multimedia text messages, refer to Chapter 9.

Your ThunderBolt Can Fly

The easiest way to make your phone fly is to keep the phone firmly in your possession and then board some type of flying craft — an airplane, a helicopter, or a UFO — and then together you can soar through the sky. It's not that difficult of a thing to do, though I've written this section to tell you some of the issues you may encounter before and during your travels aloft.

Planning for your trip

The best four-letter word you can use before leaving on a trip is *plan*. By following a plan, you can make any departure go smoothly and make the trip become more enjoyable. The planning technique works whether you're leaving for a short trip or an overseas trip or fleeing the house in terror because a huge robot is stomping down your street.

Here are some things to consider planning before you leave for a trip:

Charge the phone. The most important thing to remember before taking the ThunderBolt anywhere is to charge it. Before I leave for a long trip, I ensure that the phone has been charged overnight.

Synchronize media. Another good thing to do is synchronize media with your computer. This operation isn't so much for taking media with you, but, rather, to ensure that you have a backup of the phone's media on your computer. See Chapter 20 for synchronization information.

Get some e-reading material. Consider taking some eBooks on the road. I prefer to sit and stew over the Kindle online library before I leave, as opposed to wandering aimlessly in the airport sundry store and trying hard to focus on the good books rather than on the salty snacks. Chapter 17 covers reading eBooks on your ThunderBolt.

Save up some web pages. Another nifty thing to do is save some web pages for later reading. I usually start my day by perusing online articles and angry letters to the editor in the local paper. Because I don't have time to read that stuff before I leave, and I do have time on the plane, and I'm extremely unwilling to pay for in-flight Wi-Fi, I save my favorite websites for later reading. Here's how to save a web page when using the Internet app:

1. **Locate the link to the page you want to save.**

   It has to be a link. The phone has no obvious way to save the page you're viewing.

2. **Long-press the link.**

3. **From the pop-up menu, choose the command Save Link.**

   The web page is downloaded, saved to the MicroSD card.

Repeat these steps for each web page you want to read when offline.

To view the web page later, press the Menu soft button in the Internet app. Choose the More command and then Downloads. Select the web page from the list displayed in the Download Manager. Choose the HTML Viewer app to read the web page.

And now, the disappointing news:

A web page saved on the ThunderBolt is assigned a generic name, something like `downloadfile.htm`.

The web pages are missing their pictures; only the text and formatting are saved. Though this arrangement may not look pretty, at least you'll have some reading material on the plane — specifically, stuff you're used to reading that day, anyway.

- ✔ An app called Read it Later can be used to collect and store your favorite web pages for offline reading. It's available at the Android Market, but it's not free; it's listed for 99 cents as this book goes to press. See Chapter 18 for information on the Android Market.

- ✔ See Chapter 11 for more information on the Internet app and browsing the web on your ThunderBolt.

Entering Airplane mode

I'm not certain whether using a cell phone on an airplane interferes with the avionics, or perhaps whether airline employees (and passengers) simply enjoy being in one of the last cell-phone-free frontiers on the planet. Either way, it's common knowledge that you're forbidden from making a phone call using your cell phone while you're flying.

Though you can't make phone calls, you can use your ThunderBolt for just about anything else, including listening to music, playing games, or doing any other activity that doesn't require a cellular connection. The key to using the phone without it making a call, and risking everyone's life on the plane, is to place it into *Airplane mode.*

The most convenient way to put the ThunderBolt in Airplane mode is to press and hold the Power Lock button. From the menu, choose Airplane Mode.

The most inconvenient way to put the ThunderBolt into Airplane mode is to follow these steps:

1. **From the All Apps screen, choose the Settings icon.**
2. **Choose Wireless & Networks.**
3. **Touch the square by Airplane Mode to set the green check mark.**

   When the green check mark is visible, Airplane mode is active.

 When the phone is in Airplane mode, a special icon appears in the status area, as shown in the margin. You might also see the text *No Service* appear on the phone's locked screen.

To exit Airplane mode, repeat the steps in this section but remove the green check mark by touching the square next to Airplane Mode.

 ✔ Officially, the ThunderBolt should be powered *off* when the plane is taking off or landing. See Chapter 2 for information on turning off the phone.

✔ You can compose email while the phone is in Airplane mode. The messages aren't sent until you disable Airplane mode and connect again with a data network.

 ✔ Widgets are available that let you instantly turn Airplane mode on or off. See Chapter 18 for information about the Android Market, where you can search for Airplane mode widgets.

✔ Bluetooth networking is disabled when you place the ThunderBolt into Airplane mode, but it can be reenabled after Airplane mode is active. Even so, using Bluetooth during a flight isn't recommended on many airlines. See Chapter 19 for more information on Bluetooth.

 ✔ Many airlines now feature wireless networking onboard. You can turn on wireless networking for the ThunderBolt and use a wireless network in the air. Simply activate the phone's Wi-Fi feature, per the directions in Chapter 19, after placing the phone in Airplane mode — well, after the flight attendant tells you that it's okay to do so.

✔ Apparently, making Skype calls is possible in the air when the phone is connected to the in-flight Wi-Fi. Even so, airlines discourage the usage of Skype for making phone calls and, in fact, access to Skype may be blocked by the in-flight Wi-Fi service.

Making air travel easier

Being both a nerd and a quasi-frequent-flier, I'm becoming adept at carrying electronic junk with me when I fly. That junk includes laptops, phones, tablet computers, and various toys that don't look scary or dangerous. My experience warrants this list of high-tech travel tips:

✔ Take the ThunderBolt's AC adapter and USB cable with you. Put them in your carry-on luggage. Many airports feature USB chargers so that you can charge the phone in an airport, if you need to.

✔ At the security checkpoint, place your ThunderBolt in a bin by itself or with other electronics.

✔ Use the Calendar app to keep track of your flights. The event title serves as the airline and flight number. For the event time, use the takeoff and landing schedules. For the location, list the origin and destination airport codes. And, in the Notes field, record the flight reservation number. If you're using separate calendars (categories), specify the Travel calendar for your flight.

✔ See Chapter 17 for more information on the Calendar.

✔ Some airlines feature Android apps you can use while traveling. At the time this book goes to press, Southwest Airlines, Delta, and Continental Airlines have specific Android apps. You can use the apps to not only keep track of flights but also check in: Eventually, printed tickets will disappear and you'll merely show your "ticket" on the ThunderBolt screen, which is then scanned at the gate.

✔ Some apps you can use to organize your travel details are similar to, but more sophisticated than, using the Calendar app. Visit the Android Market and search for *travel* or *airlines* to find a host of apps.

International Calling

The old hermit didn't have a phone. He explained away that modern convenience by saying, "Why would I want a bell in my house that anyone in the world could ring?"

It's true: As long as anyone else in the world has your phone number, they can make your phone ring. Likewise, you can return the favor by ringing any phone in the world. It's not that difficult to accomplish, but international calling can be frustrating and expensive when you don't know your way around.

Dialing an international number

To make an international call on the ThunderBolt, you need to know the foreign phone number. But you also need to know how to escape the bounds of your own country's phone system to access the international phone system. It can get tricky.

The key to making a successful international call is to use the + character on your phone's dialpad. This symbol represents the *country exit code,* which must be dialed in order to flee the tyranny of your national phone system and access the international phone system.

For example, to dial Uruguay on your ThunderBolt, you dial +598 and then the number in Uruguay. The exit code consists of the plus sign (+) and then the international code for Uruguay (598).

To produce the country exit code in an international phone number, press and hold the 0 key on the ThunderBolt dialpad. You see the + symbol appear at the start of the international number. Then type the country prefix and the phone number. Touch the Call button to place the call.

✔ In most cases, dialing an international number involves a time zone difference. Before you dial, be aware of what time it is in the country or location you're calling.

✔ And keep in mind that not everyone everywhere speaks English.

✔ Dialing internationally also involves surcharges, unless your cell phone calling plan already provides for international calling.

✔ The + character is used on the ThunderBolt to represent the country exit code, which must be dialed before you can access an international number. In the United States, the exit code is 011. (In the United Kingdom, it's 00.) So, if you're using a landline to dial Mongolia from the United States, you dial 011 to escape from the United States and then 976, the country code for Mongolia. Then dial the rest of the number. You don't have to do this on the ThunderBolt, because + is always the country exit code and replaces the 011 for U.S. users.

✔ The + character isn't a number separator. When you see an international number listed as 011+20+xxxxxxx do not insert the + character in the number. Instead, dial +20 and then the rest of the number, xxxxxxx.

✔ International calls fail for a number of reasons. One of the most common is that the recipient's phone company or service blocks incoming international calls.

- Another reason that international calls fail is the zero reason: Oftentimes, you must leave out any zero in the phone number that follows the country code. So, if the country code is 673 for Brunei and the phone number starts with 012, you dial +673 for Brunei and then 12 and the rest of the number. Omit the leading zero.

- You can also send text messages to international cell phones. It works the same way as making a traditional phone call: Type the international number into the Messaging app. See Chapter 9 for more information on text messaging.

- Know which type of phone you're calling internationally — cell phone or landline. The reason is that an international call to a cell phone often involves a surcharge that doesn't apply to a landline.

Making international calls with Skype mobile

When you're counting your pennies, look into using the Skype app for making international calls. It works similarly to using the phone's cell service, but it can be much, much cheaper.

To use Skype, you need a Skype account. The best way to get one is to use your computer to visit www.skype.com, where you can set up an account and download a version of Skype for your computer.

To make an international call on Skype, you stuff your Skype account with Skype Credit. Further, you need to add a contact for the international number you plan to dial. The contact must be a Skype contact, not someone in the People app.

- You can also use Skype to text or voice-chat with other Skype users. There's no charge for this feature.

- Skype can be used to place calls to domestic phone numbers as well as to international numbers. You have to pay (with Skype Credit) to gain the ability to call regular phone numbers.

- The Skype Mobile app can be used for video chat, though the feature was disabled on the ThunderBolt when this book went to press.

The ThunderBolt Travels the Globe

My advice for traveling overseas with your phone is to first contact your cellular provider. Ask what will happen when you take the ThunderBolt to a foreign country. Someone will probably know exactly what you should expect to encounter based on where you're going. You might be able to sign up for a temporary overseas calling plan or just rent a phone that works at your destination, which could be cheaper.

You can also opt to travel with your phone but not use it as a phone while you're abroad. For example, you can use the phone's Wi-Fi service to access the Internet and not incur international roaming charges in *zloty* or *pengö*.

Though you can avoid the temptation to make a phone call, my advice is to forward all your incoming calls to another number or directly to your voice mail. That way, you don't risk taking a 6.79€ call from someone who wants to know whether you're happy with your car insurance.

A final issue is power. The ThunderBolt's AC plug can easily plug into a foreign AC adapter, which allows you to charge the phone by using that weird, 7-prong outlet in a foreign country. As long as you bring an adapter and your phone's charger, you'll be okay.

✔ A good way to keep your phone from using the cellular service abroad is to keep it in Airplane mode the entire time you're overseas.

✔ The ThunderBolt hones in on whatever cellular service is offered in whichever country you and the phone happen to be loitering. You see the foreign cellular service listed on the phone's lock screen. And, unless you want to incur data roaming charges, you should heed the advice in this section.

✔ Wi-Fi is universal, and as long as your location offers this service, you can connect the phone and pick up your email, browse the web, or do whatever other Internet activities you desire. Even if you have to pay for Wi-Fi access, I believe that you'll find it less expensive than paying a data roaming charge.

✔ Another important point to consider is security. See Chapter 22 for information on security for your ThunderBolt. Also see Chapter 24, which discusses how to find a wayward phone.

22

Personalize and Customize

*P*ersonalization is a big deal on the ThunderBolt. It's the whole theme behind HTC Sense, which is the graphical user interface you use on the phone's touchscreen. In fact, personalization is such a big deal that you can find the Personalize button at the bottom of every Home screen. The Personalize command is also found at the top of the Settings screen.

Oh, and you have this chapter, which covers the entire personalization topic. The idea is to make the phone your own by customizing its appearance to just the way you like.

Personalize This, Personalize That

When you touch the Personalize button (shown in the margin and found at the bottom of every Home screen), you see the Personalize screen, shown in Figure 22-1. It's chock-full of things you can do to truly make your phone your own.

Put things on the Home screen.

Change the way things look.

Change the way things sound.

Figure 22-1: The Personalize screen.

You can also access the Personalize screen from the phone's main control center — the Settings screen: From the Home screen, press the Menu soft button and choose Settings and then Personalize. (Long-pressing a blank part of the Home screen also summons the Personalize screen.)

The remainder of this chapter discusses the various things you can do on the Personalize screen, which helps you make your ThunderBolt phone unique and, well, personal.

✔ Personalization is optional. If you're happy with the way your phone looks, you're good.

✔ The best things to personalize are the shortcut icons and widgets on the Home screen. See the later section "Home Screen Improvement" for details.

I've Got You Under My Skin

Despite its high-tech, composite design, the ThunderBolt indeed has *skin*. No, not the kind of skin that can itch, but rather a skin that can change the way the phone looks. This section covers changing the skin and other ways you can personalize the Home screen's appearance.

Changing scenes and skins

Two items on the Personalization screen deal with the way your phone presents its overall appearance: Scene and Skin.

Scene: The Scene is a combination of wallpaper, or the image on the Home screen background, and app shortcut and widget placement.

Skin: The Skin refers to the look and feel of the graphical interface. It describes how the lock screen's lock bar looks, how the icons at the bottom of every Home screen appear, and how every window or screen is shaded and colored.

To choose a new scene or skin, choose one from the Scene or Skin categories found on the Personalize screen, as shown earlier, in Figure 22-1. Swipe the items left and right to peruse the variety. Touch the Apply button to select a scene or skin.

✔ The Home screen background, or *wallpaper,* can be set separately from the scene. See the next section.

✔ Scenes and skins can also be viewed in List mode: Press the Menu soft button and choose the List command. To switch back to non-List mode, press the Menu soft button and choose the Panel command.

✔ If you don't like the variety of scenes, press the Menu soft button while viewing the Scenes screen and choose the Add command. A new, empty scene is created, which you can populate with widgets and icons as you see fit.

✔ You cannot create your own skin using the tools available on the ThunderBolt.

✔ The ThunderBolt ships with the Verizon Scene and HTC Skin selected.

Hanging new wallpaper

The Home screen has two types of backgrounds, or *wallpapers:* traditional and live. A *live* wallpaper is animated. A not-so-live *(traditional)* wallpaper can be any image, such as a picture you've taken and stored in the Gallery.

To set a new wallpaper for the Home screen, obey these steps:

1. **Long-press the Home screen.**

   The Personalize screen appears, as shown earlier, in Figure 22-1.

2. **Choose Wallpaper.**

   You have three options for choosing wallpaper:

   HTC Wallpapers: Choose a wallpaper from a range of images preinstalled on the ThunderBolt.

   Live Wallpapers: Choose an animated or interactive wallpaper from a list.

   Gallery: Choose a still image from those you've taken, stored in the Gallery.

3. **Select a wallpaper option type.**

4. **Choose the wallpaper you want from the list.**

   For HTC wallpapers and live wallpapers, scroll left and right and stop at the wallpaper you like.

   For the Gallery option, you see a preview of the wallpaper where you can select and crop part of the image.

   For certain live wallpapers, a Settings button may appear. The settings let you customize certain aspects of the interactive wallpaper.

5. **Touch the Preview button to see how the wallpaper will look or, for a Gallery wallpaper, touch the Save button and you're done.**

6. **Touch the Apply button to confirm your selection or press the Back soft button to chicken out.**

   The new wallpaper takes over the Home screen.

Live wallpaper features some form of animation, which can often be interactive. Otherwise, the wallpaper image scrolls slightly as you swipe from one Home screen panel to another.

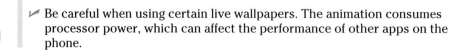

- ✒ The Zedge app has some interesting wallpaper available. It's an über-repository of wallpaper images, collected from Android users all over the world. Check out Zedge at the Android Market; see Chapter 18.

- ✒ See Chapter 15 for more information about the Gallery, including information on how cropping an image works.

- ✒ Be careful when using certain live wallpapers. The animation consumes processor power, which can affect the performance of other apps on the phone.

Rearranging Home screens

There are seven total Home screens on the ThunderBolt. You cannot add or remove Home screens, but you can rearrange them: While viewing the main Home screen, press the Home soft button. You see Leap view, shown in Figure 22-2.

Insertion pointer shows where
the Home screen will be moved.

Long-press a Home screen
to pick it up and move it.

The center Home screen is the main Home screen.

Figure 22-2: Manipulating Home screens.

You can use Leap view to instantly switch to another Home screen by touching it, but you can also drag around the Home screens with your finger, rearranging their order, as illustrated in Figure 22-2.

Press the Back or Home soft buttons when you're done rearranging.

✓ Any Home screen can be moved to any other position. The numbers 1 through 7 in Figure 22-2 represent the Home screen positions from the far left Home screen (number 1) through the far right Home screen (number 7). Home screen number 4 (in the center) is the main Home screen.

✔ The main Home screen is the one you return to when you press the Home soft button.

✔ The phone vibrates a tad as you "pick up" a Home screen. After vibrating, the Home screen can be moved.

✔ There's no way to undo a Home screen rearranging operation, other than to move things manually back to the way they were.

✔ Refer to Chapter 3 for information on using Leap view.

Home Screen Improvement

The stars may be fixed in the heavens, but the icons and doodads you see on the ThunderBolt's Home screen are temporary and variable. That's because you can add new items to the Home screen, move existing items, and even kill things off.

Putting favorite apps on the Home screen

The first thing I did on my ThunderBolt was to place my favorite apps on the Home screen. Here's how it works:

1. **View the Home screen on which you want to add an app shortcut.**

   Ensure that the Home screen has room for the app icon.

2. **Press the Personalize button.**

   The Personalize screen appears, as shown in Figure 22-1.

3. **Choose App.**

4. **Select an app from the list to add to the Home screen.**

   The app appears on the Home screen you chose in Step 1.

If there's no room for the app, you see a warning on the screen explaining as much. Swipe to another, empty Home screen and then touch the app's icon to attach it to that Home screen.

Another, often quicker way to add an app shortcut icon to the Home screen is to summon the All Apps screen and long-press an app. The phone vibrates for a tick and then the app — which remains stuck under your finger — appears on the Home screen. Drag the app icon around, if you like. Release your finger to drop the app on the Home screen.

Too many apps on the Home screen!

The ThunderBolt features seven Home screens on which you can affix your favorite apps. If you find yourself with more apps than will fit on a screen — or on all the screens — several solutions are available to you.

The first solution is provided by the Android operating system, in the form of *folders*. The folders work similarly to folders on your computer in that you can stuff a folder full of app icons, keeping them all in one handy place. For example, I typically use a folder for my games because it keeps them in one place and I name the folder *Work* just to throw off anyone who thinks I play with my phone more than I use it for work.

To create a folder, choose the Folder command from the Personalize screen, shown in Figure 22-1. Choose the New Folder command. You move apps into the folder by dragging their icons to the folder's icon. Open the folder to view its contents, start an app, or drag an app out of the folder and back to the Home screen.

You can change a folder's name by long-pressing the title when the folder is open.

The second solution is to use the Apps Organizer app. It's a bit more sophisticated than using folders, and it involves more work: You have to tag your apps into certain categories, such as games, news, or shopping. Apps Organizer builds folders for the apps based on their tags. Check out Apps Organizer at the Android Market; see Chapter 18.

Also see the later section "Moving and removing icons and widgets" for moving app icons on the Home screen.

- ✔ Apps aren't moved to the Home screen: What you see is a copy, a shortcut. You can still find the app on the All Apps screen, but now the app is — more conveniently — available on the Home screen.

- ✔ Keep your favorite apps, those you use most often, on the Home screen.

- ✔ Icons on the Home screen are aligned to a grid. You can't stuff more icons on the Home screen than will fit in the grid, so when a Home screen is full of icons (or widgets), use another Home screen.

Working with widgets

A *widget* works like a tiny, interactive or informative window, often providing a gateway into another app on the ThunderBolt. Just as you can add apps to the Home screen, you can add widgets.

The ThunderBolt comes with a smattering of widgets already affixed to the Home screen, possibly just to show you how they can be used. You can place even more widgets on the Home screen by following these steps:

1. **Locate a Home screen with plenty of free space for the widget.**

2. **Long-press the Home screen to summon the Personalize screen.**

3. **Choose Widget.**

4. **From the list, choose the widget you want to add.**

5. **Optionally, make additional selections, depending on the widget.**

   The widget is plopped on the Home screen.

The variety of available widgets depends on the applications you have installed. Some applications come with widgets, some don't. Some widgets come independently of any application.

 ✔ More widgets are available at the Android Market. See Chapter 18.

 ✔ You cannot install a widget when the Home screen has no room for it. Choose another panel, or remove icons or widgets to make room.

 ✔ To remove a widget, see the later section "Moving and removing icons and widgets."

Making some shortcuts

A *shortcut* is a doodad you can place on the Home screen that's neither an app nor a widget. Instead, a shortcut is a handy way to get at a feature or an informational tidbit stored in the ThunderBolt without having to endure complex gyrations.

For example, I have a shortcut on my Home screen that uses the Maps app Navigation feature to help me return to my house. I put the shortcut there in case of a zombie attack so that I can return home to my hardy stock of power tools.

To add a shortcut, long-press the Home screen and choose the Shortcuts command from the Personalize screen (refer to Figure 22-1). What happens next depends on which shortcut you choose.

For example, when you choose a bookmark, you add a web page bookmark to the Home screen. Touch the shortcut to open the Internet app and pluck the web page from your list of bookmarks.

Choose a Person shortcut to display contact information for a specific contact in the phone's address book.

A nerdy shortcut to add is the Settings shortcut. After choosing this item, you can select from a number of on-off options or status items that can appear on the Home screen as widgets.

 The Any Cut app is useful for creating certain shortcuts that the ThunderBolt cannot create by itself, such as a shortcut to direct-dial a contact. Check out Any Cut at the Android Market; see Chapter 18.

Moving and removing icons and widgets

There's no reason for you to tolerate or accept any arrangement of icons and widgets on the Home screen, especially those items preset by the manufacturer. At any time, you're free to move things about, and even delete unwanted items. Truly, you are lord of the Home screen.

To move an icon or a widget, long-press it. Eventually, the icon seems to lift and break free, as shown in Figure 22-3.

Long-press an icon.

Drag the icon to the trash.

Figure 22-3: Moving an icon about.

You can drag a free icon to another position on the Home screen or to another Home screen panel, or you can drag it to the Remove (trash can) icon that appears at the bottom of the Home screen (refer to Figure 22-3).

Widgets can be moved around or deleted in the same manner as icons.

✏ Removing a Home screen icon or widget doesn't uninstall the application, which is still found on the All Apps screen. Apps and widgets can be added to the Home screen again at any time, as described earlier in this chapter.

✏ When an icon hovers over the Remove button, ready to be deleted, its color changes to red.

✏ See Chapter 18 for information on uninstalling applications.

ThunderBolt Locks

The ThunderBolt features a basic lock screen: Simply slide down the Lock bar and the phone is unlocked and ready to use. If you prefer to have a lock that's not so easy to pick, you can choose from one of three different types of lock screens: Pattern, PIN, and Password, as described in this section.

Locating the locks

Lock screen security is set on the Set Screen Lock screen. Here's how to get there:

мєпu

1. **From the Home screen, press the Menu soft button.**

2. **Choose Settings.**

3. **Choose Security.**

   The Security screen appears.

4. **Choose Set Up Screen Lock.**

5. **If a screen lock is already set, you must trace the pattern or input the PIN or password to continue.**

   The Screen Unlock Security screen lists four types of locks:

   None: Rather than no lock, the phone simply uses the standard locking screen: Slide down the lock bar and the phone is unlocked. Choosing this option disables the other three options.

   Pattern: To unlock the phone, you must trace a pattern on the touchscreen.

   PIN: The phone is unlocked by typing a personal identification number (PIN).

   Password: You must type a password to unlock the ThunderBolt.

6. **To set a lock, refer to the following sections, or to remove any existing lock, choose None.**

When you choose None to remove another type of lock, you're asked to unlock the phone one more time, by tracing the pattern or typing the PIN or password to remove that type of lock.

- ✔ The security you add affects the way you turn on and wake up your ThunderBolt. See Chapter 2 for details.

- ✔ A specific lock (pattern, PIN, or password) doesn't show up when you've just "slept" the phone. A timeout value set on the Security screen determines when the lock appears: Choose Lock Phone After and then set the time in minutes. The ThunderBolt is preconfigured to use the pattern, PIN, or password lock after 15 minutes of inactivity.

- ✔ The security lock can be overridden when USB debugging is enabled. Unless you're writing software for the ThunderBolt, however, odds are good that you'll never have USB debugging turned on.

Creating an unlock pattern

One of the most common ways to lock the phone is to apply an *unlock pattern:* The pattern must be traced exactly as it was created in order to unlock the device and gain access to your apps and other ThunderBolt features.

To set the unlock pattern, summon the Screen Unlock Security screen as described in the preceding section. Choose Pattern.

If you've not yet set a pattern lock, you may see a text screen describing the process; touch the Next button to work the dreary directions.

Trace an unlock pattern, as shown in Figure 22-4. You can trace over the dots in any order, but you can trace over a dot only once. The pattern must cover at least four dots.

Touch the Continue button and redraw the pattern again, just to prove to the doubtful ThunderBolt that you know the pattern.

Touch the Confirm button and the pattern lock is set.

Ensure that a check mark appears by Use Visible Pattern on the Security screen. The check mark ensures that the pattern shows up. For even more security, you can disable the option, though you have to be sure to remember how — and where — the pattern goes.

- ✔ To remove the pattern lock, set None as the type of lock, as described in the preceding section.

- ✔ The pattern lock can start at any dot, not necessarily the upper right dot, as shown in Figure 22-4.

- ✔ The unlock pattern can be as simple or complex as you like. I'm a big fan of simple.

- ✔ Wash your hands! Smudge marks on the display can betray your pattern.

1. Start here.

2. Trace pattern.

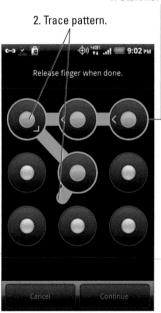

Figure 22-4: Setting an unlock pattern.

Setting a PIN

I suppose that using a PIN, or *personal identification number*, is more left-brained than using a pattern lock. What's yet another number to memorize?

A *PIN lock* is a code between 4 and 16 numbers long. It contains only numbers, 0 through 9. To set a PIN lock for your ThunderBolt, follow the directions in the earlier section "Locating the locks" to reach the Set Screen Lock screen. Choose PIN from the list of locks.

Input your PIN twice to confirm that you know it. The next time you need to unlock your phone, type your PIN on the keypad and touch the OK button to proceed.

- ✔ Refer to Figure 2-2 (in Chapter 2) for an image of what the PIN lock looks like when you unlock your ThunderBolt.

✓ To disable the PIN, reset the ThunderBolt security level to None, as described in the section "Locating the locks," earlier in this chapter.

Applying a password

The most secure way to lock the ThunderBolt is to apply a full-on password. Unlike a PIN (refer to the preceding section), a *password* can contain numbers, symbols, and both upper- and lowercase letters.

Set a password by choosing Password from the Screen Unlock Security screen; refer to the earlier section "Locating the locks" for information on getting to that screen.

The password you create must be at least four characters long. It can contain letters, numbers, and other characters available on the keyboard. Longer passwords are more secure, but easier to mistype.

You need to type the password twice to set things up, which confirms to the doubting phone that you know and will hopefully remember the password.

The ThunderBolt prompts you to type the password whenever you unlock the phone, as discussed in Chapter 2. You also need to type the password whenever you try to change or remove the screen lock, as discussed in the section "Locating the locks," earlier in this chapter.

ThunderBolt Tunes and Tweaks

Adjusting things on your phone can be a full-time obsession, which I suppose is the whole point behind the Personalization screen (refer to Figure 22-1). Of all the things you can adjust, the more vital items are covered in this section.

Assigning common sounds

The Personalize Sound section of the Personalize screen lists three sound-generating items you can set: Ringtone, Notification Sound, and the sound that plays for the Clock app's alarm.

Choose Ringtone to set the phone's ringtone. Specific directions are found in Chapter 6.

Choose the Notification Sound item to see a list of apps and events that generate notifications on the ThunderBolt. Choose an item from the list to select a sound, or ringtone, for that event or app.

Choose the Close item to set the sound that plays for the alarm in the Clock app, as described in Chapter 17.

All your sound choices can be saved at one time as a sound set by choosing the Sound Set item: Touch the New Sound Set button and give your sound set a name.

You can use sound sets to quickly change all the sounds your ThunderBolt makes, by choosing only one item.

Controlling the noise

General noisiness on the ThunderBolt is handled by the Sound item on the Settings screen: From the Home screen, press the Menu soft button and choose Settings, and then choose Sound to see the Sound Settings screen.

You should consider a few interesting and useful items using the Sound Settings screen:

Volume: Though you can set the ThunderBolt volume using the Volume buttons on the side of the gizmo, the Volume command on the Sound Settings screen lets you set the volume for a number of sound events, as shown in Figure 22-5. Table 22-1 describes what the various volume items control.

Figure 22-5: Various volume settings.

For example, if you want the alarms to be loud, and all those notification sounds to be rather mute, adjust the sliders (refer to Figure 22-5) accordingly.

Vibrate: Place a check mark by the Vibrate item to have the phone vibrate on incoming calls and notifications.

Quiet Ring On Pickup: After ringing for an incoming call, the phone immediately grows quieter after you move it.

Pocket Mode: The ThunderBolt uses its proximity sensor to determine when it's in a pocket or purse and then increases the volume.

Vibrate Feedback: When set, this option directs the phone to provide a slight vibration as you manipulate items on the touchscreen.

Of course, every item on the Sound Settings screen has a purpose. Feel free to set and use the items to configure the ThunderBolt to make as much or little noise, and the kinds of noise, you prefer.

Table 22-1	Various Noisy Things
Volume Setting	*Sounds It Controls*
Ringtone	Incoming calls
Media	Music, video, YouTube, Internet, game, and others
Alarm	Warnings set by the Clock app, and other alerts
Notification	New notifications, such as new email messages and calendar appointments, and other notification-generating apps and events

Adjusting screen brightness

Probably the key thing you want to adjust visually on the ThunderBolt is screen brightness. Follow these steps:

1. **At the Home screen, press the Menu soft button and choose the Settings command.**

2. **Choose Display.**

3. **Choose Brightness.**

4. **Adjust the slider to make the screen brighter or less bright.**

   The touchscreen adjusts its intensity as you move the slider, giving you instant feedback.

5. **Touch the OK button.**

If you place a check mark by the Automatic Brightness setting (after Step 3), the ThunderBolt adjusts the screen brightness based on the general luminosity of your location. I find this setting useful when I use the phone in my car, because the ThunderBolt screen isn't too bright at night or too dim during the day.

Setting the screen timeout

To save power, the ThunderBolt dims the touchscreen display after a given period of inactivity (or indifference), and then it turns off the screen. To adjust the timeout value, follow these steps:

1. **At the Home screen, press the Menu soft button.**

2. **Choose Settings.**

3. **Choose Display.**

4. **Choose Screen Timeout.**

   Select a new timeout value from the Screen Timeout menu. The new timeout value is set instantly.

The timeout values range from 15 seconds to Never Turn Off, which is an option I don't recommend. (The phone's battery doesn't like the Never Turn Off setting.)

The ThunderBolt is preset with a screen timeout value of one minute.

23

All Shiny and Spiffy

My experience is that most people greet the welcome and necessary topic of routine maintenance with a hearty "Why bother?" In a society weaned on disposable products, the whole topic of maintenance seems old fashioned. I feel differently, of course, and have grown fond of keeping my electronic doodads in top working order. And, if they don't work properly, at least they look bright and shiny.

Maintaining your ThunderBolt is a cinch. Honestly, you have to do a few necessary maintenance things on a phone, but I do them anyway. Beyond maintenance, this chapter covers battery-saving information, helpful tips, and a spiffy smattering of Q&A.

The Maintenance Thing

There's no oil to change, no tires to rotate. You don't have to take your ThunderBolt into the Phone Store every 20,000 minutes to get a 15-point inspection. Happily, there's no need to winterize your ThunderBolt. When it comes to maintenance, you have only a smattering of things to consider, as offered in this section.

Cleaning the phone

Unlike your glasses, you probably keep your ThunderBolt clean without even thinking about it. Maybe you wipe the touchscreen with your sleeve, or perhaps merely shoving the phone in and out of your pocket gives it that nice, buffed polish you'd appreciate for your dress shoes. Either way, keeping a phone clean isn't much to worry about.

I recommend two items for keeping your ThunderBolt clean. The first is a *microfiber cloth,* which can be found at any computer- or office-supply store. The cloth is ideal for cleaning the touchscreen as well as the rest of the phone (and your glasses).

The second item is a *screen protector* for the touchscreen. I used to question these transparent, clingy things, but now I'm a believer. If you tire of constantly wiping off the touchscreen, pick up some screen covers. Ensure that you get covers specifically designed for the ThunderBolt.

Never use any liquid to clean the touchscreen — especially ammonia or alcohol. These chemicals can damage the touchscreen, rendering it unable to read your input. Further, they can smudge the display, making it more difficult to see.

Backing up your stuff

A *backup* is a safety copy of information. For your ThunderBolt, the backup copy includes contact information, music, photos, videos, and apps you've installed, plus any settings you've made to customize your phone. Copying this information to another source is one way to keep the information safe, in case anything happens to your ThunderBolt.

Yes, a backup is a good thing. Lamentably, there's no universal method of backing up the stuff on your ThunderBolt.

Your Google account information is backed up automatically. This information includes your Contacts list, Gmail inbox, and Calendar app appointments. Because the information automatically syncs with the Internet, a backup is always present.

To confirm that your Google account information is being backed up, heed these steps:

1. **From the Home screen, press the Menu soft button.**
2. **Choose Settings.**
3. **Choose Accounts and Sync.**

4. **Ensure that both the Background Data and Auto-Sync items have green check marks by them.**

5. **Touch the green Sync button by your Google account name.**

6. **Ensure that check marks appear by every item in the list.**

   On my ThunderBolt, the list includes Books, Contacts, Gmail, and Calendar.

7. **Press the Back soft button.**

8. **Optionally, ensure that your other accounts are being synchronized as well.**

   When you have more than one Google account synchronized with your ThunderBolt, repeat Steps 1 through 6 for every account. Ditto for your Flickr and Mail accounts.

9. **Press the Back soft button to return to the main Settings screen.**

10. **Choose Privacy.**

11. **Ensure that a check mark appears by the item Back Up My Settings.**

    You should see a check mark there. If not, touch the square to add one.

When you have a PC, you can use the HTC Sync program to synchronize additional information for your phone. See Chapter 20 for information on using HTC Sync.

Applying system updates

Every year or so, Google comes out with a new version of Android, the ThunderBolt's operating system. A short time afterward, an operating system update might become available for your phone.

Upgrading the operating system is an automatic thing; you're alerted when the update is available. You usually have three options for installing a system update:

- ✔ Install Now
- ✔ Install Later
- ✔ More Info

My advice is to choose Install Now and get it over with — unless you're doing something urgent, in which case you can put off the update until later by choosing Install Later.

✔ You can manually check for updates: From the Settings screen, choose the Software Update command and then choose Check New. When your system is up-to-date, the screen tells you so. Otherwise, you find directions for updating the Android operating system.

✔ Non-Android system updates might also be issued. For example, HTC may send out an update to the ThunderBolt's guts or to the HTC Sense user interface. This type of update is often called a *firmware* update. As with Android updates, my advice is to accept and install all firmware updates.

The Happy Battery

In all your phone's guts, perhaps the one part you'll most likely pay attention to is the battery. Keeping the battery charged, and making the charge last as long as you can, is a vital part of using the phone. After all, when the battery power is gone, the ThunderBolt is little more than an expensive nutcracker.

Checking battery status

When you use the phone a lot, you'll probably look at the battery meter more than anything else. You'll especially notice the icon when it turns a delightful yellow color, which means that the battery is beginning to get desperate for a charge.

Figure 23-1 describes the various types of battery status icons available on the ThunderBolt.

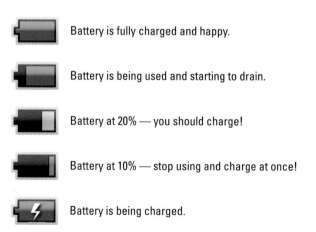

Figure 23-1: Battery status icons.

You might also see the icon for a dead battery, but for some reason I can't get my ThunderBolt to turn on and display that icon.

When you find the teensy battery icons (refer to Figure 23-1) too vague, you can check the specific battery level by following these steps:

1. **From the Home screen, press the Menu soft button.**

2. **Choose Settings.**

3. **Choose About Phone.**

4. **Choose Battery.**

The top two items on the Battery screen offer the detailed information you're after:

Battery Status: This setting explains what's going on with the battery. It might say Full or Charging, or you might see other text, depending on how desperate the ThunderBolt is for power.

Battery Level: This setting reveals a bar that graphically shows battery power. A full, green bar means that the battery is charged at 100 percent. The green bar gets shorter as battery power is used.

Battery Use: This command displays the items that consume battery power, as described in the next section.

You don't have to keep checking the Battery screen as you use your ThunderBolt. The battery icon on the status bar (refer to Figure 23-1) is your best clue for battery consumption. Also, the ThunderBolt pops up various warnings when the battery power drains to 20 and 10 percent, respectively. The 10 percent warning is the serious one: Plug the phone into a power source or turn it off at once!

✒ Heed those low-battery warnings!

✒ When the battery level is too low, the ThunderBolt shuts itself off.

✒ The ThunderBolt battery gets beaten up by a lot of things, especially the 4G LTE network, so the phone may seem to just sip battery power for a while. But if you make heavy-duty cellular data transfers over the 4G LTE network, the battery meter starts to *move*.

✒ The best way to deal with low battery power is to connect the phone to a power source: Either plug it into a wall socket or connect it to a computer by using a USB cable. The ThunderBolt charges itself immediately; plus, you can use the device while it's charging.

> ✔ You don't have to fully charge the ThunderBolt to use it. If you have only 20 minutes to charge and you get only a 70 percent battery level, that's great. Well, it's not great, but it's far better than a 20 percent battery level.

> ✔ Battery percentage values are best-guess estimates. Even if you don't use the phone, the battery drains faster as the power level decreases. Don't view the battery meter as a reliable measure of remaining power.

Finding power hogs

Not everything the ThunderBolt does thirsts for power. To find out which items consume most of your phone's precious battery life, you can visit the handy, informative screen shown in Figure 23-2. To get to that screen, follow these steps:

Figure 23-2: Things that drain the battery.

1. **At the Home screen, press the Menu soft button.**

2. **Choose Settings.**

3. **Choose About Phone.**

4. **Choose Battery.**

5. **Choose Battery Use.**

   You see a screen similar to the one shown in Figure 23-2.

6. **Touch an item in the list to see more information about how it affects battery use.**

   Occasionally, you find additional options on the Battery Use Details screen, which can help you reduce battery consumption.

One of the biggest power hogs is the display. If you choose the Display item on the Since Unplugged screen (refer to Figure 23-2), you see more details on how the display is draining battery, with perhaps a suggestion to reduce screen brightness. The Display Settings button helps you to go instantly to the screen, where you can make additional adjustments.

The number and variety of items listed on the Battery Use Details screen depend on what you've been doing between charges and how many apps you're using.

Saving power with Power Saver

Don't get your hopes up. The ThunderBolt features a power saver, a type of power management, but it kicks in only when the battery level gets precariously low. You can visit the Power Saver screen by following these steps:

1. **At the Home screen, press the Menu soft button.**

2. **Choose Settings.**

3. **Choose Power.**

4. **Ensure that the Enable Power Saver item has a green check mark by it.**

   The check mark indicates that the Power Saver is active and kicks in at the time indicated by the Turn Power Saver On At setting.

5. **Choose the Power Saver Settings item to peruse the actions the power saver takes when it kicks in.**

Though I think the Power Saver feature is a keen idea, it's akin to putting a lifeboat on your yacht. A better feature for the ThunderBolt would be power management software, which you can use to control energy-consuming features at all times, not just when the juice level gets low.

TIP

Going battery nuts

Battery beefiness is measured in amperes. As is the case with most things you want, the more amperes in a battery, the longer the battery can power your gizmos.

For the ThunderBolt, the battery that ships with the phone is a 1400 mAh battery. The *mAh* stands for *milliampere-hours*, though it's also the noise my cat makes when he wants outside. Given all you do with your phone, often those 1400 mAhs go away pretty darn fast. The solution? Get a beefier battery.

Because the ThunderBolt battery is removable, it's possible to get a larger battery, which means, technically, a battery with the same dimensions but more milliampere-hours. Several outfits are available that sell ThunderBolt-compatible 2750 mAh batteries — nearly twice the oomph in the same package size.

The only drawback to getting a beefier, more ampere-packed battery for the ThunderBolt is that it makes the already heavy phone even heavier. The battery also adds a bit to the phone's girth, but it remains a viable solution.

You can find beefier batteries for the ThunderBolt online and possibly at the store where you purchased your phone. Additional, standard batteries can be purchased as well, along with multiple-battery charging docks.

Yeah, it's completely possible to go nuts over your phone's battery.

Making the battery last longer

Unless you plan on never turning on the phone, the battery drains over time. How quickly it drains depends on how much you use the phone. Some things use more power than others. For example, the touchscreen uses a lot of power, which is why the phone puts the touchscreen to sleep when you're on a call.

Beyond not using the phone, you can try a few things to help prolong battery life. Here's my list:

Turn off vibration options: The phone's vibration is caused by a teensy motor. Though you don't see much battery savings by disabling the vibration options, it's better than no savings. See Chapter 22 for information on the ThunderBolt sound settings, where you find the vibration options.

Lower the volume: Additionally, consider lowering the volume of the various noises the ThunderBolt makes, especially notifications. Information on setting volume options is also found in Chapter 22.

Dim the screen: If you look at Figure 23-2 (earlier in this chapter), you see that the display sucks down quite a lot of battery power. Though a dim screen can be more difficult to see, especially outdoors, it definitely saves on battery life.

Turn off Bluetooth: When you're not using Bluetooth, turn it off. See Chapter 19 for information on Bluetooth, though you can turn it off easily from the quick actions at the top of the notification panel.

Though the ThunderBolt has no power management tool, you can find plenty available at the Android Market. One that I can recommend is Juice Defender. It allows you to create a power management plan for your phone, similar to the power-savings plans you can implement (but probably don't) on your computer.

Juice Defender comes in three flavors: Battery Saver, Plus, and Ultimate, depending on how intricately you want to tune your phone's power management. The QR code for the Battery Saver version of the app is shown in the margin.

Help!

Perhaps the most useful four-letter word you can apply in dire situations is *help*. It may not be the first four-letter word that comes out of your mouth, but it's probably the most effective, especially when you can find help for your ThunderBolt right here in this section.

Fixing problems

Here are some typical problems you may encounter on the ThunderBolt, and my suggestions for a solution:

General trouble: For just about any problem or minor quirk, consider restarting the ThunderBolt: Press and hold the Power Lock button. From the menu that appears, choose the Restart command. This procedure will most likely fix a majority of the annoying and quirky problems you encounter.

Check the cellular data connection: As you move about, the cellular signal can change. In fact, you may observe the icon on the status bar change from 4G to 3G to 1X, indicating that you're in an area that doesn't offer full-strength cellular data service. And, sometimes the signal goes away altogether.

My advice for random signal weirdness is to wait. Oftentimes, the signal returns after a few minutes. If it doesn't, the cellular data network might be down or you may just be in an area with lousy service. Consider changing your location.

Check the Wi-Fi connection: For Wi-Fi connections, you have to ensure that the Wi-Fi is set up properly and working. This process usually involves pestering the person who configured the Wi-Fi signal or made it available, such as the cheerful person with the bad haircut who serves you coffee.

Music begins to play while you're on the phone: It's possible to accidentally start playing music while you're on a phone call. Or, maybe the music was playing when a call came in and doesn't stop automatically. Either way, the quickest way to stop a song from playing is to pull down the notifications and choose the playing song from the list. Touch the Pause button to pause the music.

The MicroSD card is busy: Most often, the MicroSD card is busy because you've mounted it on your computer's storage system. To "unbusy" the MicroSD card, unmount the MicroSD card by setting the USB connection to Charge Only. See Chapter 20.

When the MicroSD card remains busy, consider restarting the ThunderBolt, as described earlier in this section.

An app has run amok: Sometimes, apps that misbehave let you know. You see a warning on the screen announcing the app's stubborn disposition. Touch the Force Stop button to shut down the errant app.

When you see no warning or an app appears to be unduly obstinate, you can shut 'er down the manual way, by following these steps:

1. **From the All Apps screen, choose the Settings icon.**

2. **Choose Applications.**

3. **Choose Manage Applications.**

   If you choose the Running tab from the top of the Manage Applications screen, you see a list of only the apps that are running.

4. **Choose the app that's causing you distress.**

   For example, a program doesn't start or says that it is busy or has another issue.

5. **Touch the Force Stop button.**

   The program stops.

After stopping the program, try opening it again to see whether it works. If the program continues to run amok, contact its developer: Open the Market app, press the Menu soft button, and choose My Apps. Choose the app you're having trouble with, scroll to the bottom of the app's main screen, and choose Send Email to Developer. Send the developer a message describing the problem.

Reset the phone's software: When all else fails, you can do the drastic thing and reset all phone software, essentially returning it to the state it was in when it first arrived. Obviously, you need not perform this step lightly. In fact, consider finding support (see the next section) before you start:

1. **From the Home screen, touch the Menu soft button.**

2. **Choose Settings.**

3. **Choose SD & Phone Storage.**

4. **Choose Factory Data Reset.**

5. **Touch the Reset Phone button.**

6. **Touch the Erase Everything button to confirm.**

   All the information you've set or stored on the ThunderBolt is purged.

Again, do not follow these steps unless you're certain that they will fix the problem or you're under orders to do so from someone in Tech Support.

A factory data reset doesn't erase photos or music you've installed on the MicroSD card. It does, however, erase just about everything else.

Seeking support

Support is available for your ThunderBolt by dialing the cell phone's support number, 611. Dialing this number puts you in touch with Verizon, where you can either interact with its robot or eventually get in touch with a live human being who's eager to help you with various cell phone issues.

On the Internet, you can find support at these websites:

```
www.htc.com/www/support.aspx
```

```
http://market.android.com/support
```

```
http://support.vzw.com/clc
```

ThunderBolt Q&A

I enjoy finding a Q&A section in a book. The section is an effective way to present problems and solutions, but also occasionally hits upon a question you might have about your phone or presents a solution to a problem that's vexing you.

"My Home screen app shortcuts disappear!"

Your app shortcuts should stay affixed to the Home screen, even after you've turned the phone off and then on again. The only time the shortcuts disappear is when the apps are installed on the MicroSD card rather than in the ThunderBolt's main memory. In that situation, apps you've placed on the Home screen disappear whenever you mount the MicroSD card into your computer's storage system.

The solution is to move the app from the MicroSD card to the phone's main storage. See Chapter 18 for information.

"The touchscreen doesn't work!"

A touchscreen, such as the one used on the ThunderBolt, requires a human finger, or a similar dexterous appendage, for proper interaction. The phone interprets the static potential between the human finger and the device to determine where the touchscreen is being touched.

You cannot use the touchscreen when you're wearing gloves, unless they're specially designed gloves that claim to work on touchscreens. Or, they could be regular gloves that have been treated with special, scientific, touchscreen-transforming gel. I'm serious — this stuff actually exists.

The touchscreen might also fail when the battery power is low or when the ThunderBolt has been physically damaged.

"The sound is messed up"

Maybe you don't hear the touch-tones when you dial a number, or maybe the phone vibrates instead of ringing. This type of problem can come about because of an app you've installed on the phone. Some apps have been known to hijack the ThunderBolt's sound system or the vibration feature, though the hijacking isn't intentional.

The quick solution is to uninstall the app that's causing the problem. This solution might not be practical because you may be unable to remember which app you installed recently or pinpoint which app caused the issue.

Sometimes, restarting the phone fixes the problem. The desperate solution is to reset the phone's software, as described in the section "Fixing problems," earlier in this chapter.

"The onscreen keyboard is too small!"

You can rotate the phone to landscape orientation to see a larger onscreen keyboard. Not every app may feature a landscape orientation keyboard. When one does, you find typing on the wider onscreen keyboard much easier than normal.

"The battery doesn't charge"

Start from the source: Is the wall socket providing power? Is the cord plugged in? The cable may be damaged, so try another cable.

When charging from a USB port on a computer, ensure that the computer is turned on. Most computers don't provide USB power when they're turned off.

"The phone gets so hot that it turns itself off!"

Yikes! An overheating gadget can be a nasty problem. Judge how hot the ThunderBolt is by seeing whether you can hold it in your hand: When it's too hot to hold, it's too hot. If you're using the ThunderBolt to cook an egg, it's too hot.

Turn off the ThunderBolt and let the battery cool.

If the overheating problem continues, have the ThunderBolt looked at for potential repair. The battery might need to be replaced, and, as far as I can tell, there's no way to remove and replace the ThunderBolt battery by yourself.

Do not continue to use any gizmo that's too hot! The heat damages the electronics. It can also start a fire.

"The phone doesn't do Landscape mode!"

Not every app takes advantage of the ThunderBolt's ability to orient itself in Landscape mode, or even Upside-Down mode. For example, many games set their orientations one way and refuse to change, no matter how you hold the phone. So, just because the app doesn't go into Landscape mode doesn't mean that anything is broken.

Confirm that the Auto Rotate Screen option is on: From the Settings screen, choose Display and ensure that there's a check mark by the Auto Rotate Screen option. When the option is enabled, the ThunderBolt screen should auto-rotate. If not, the app you're using simply doesn't rotate.

Part VI
The Part of Tens

The 5th Wave — By Rich Tennant

"Well, here's what happened—I forgot to put it on my calendar."

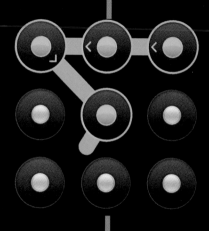

In this part . . .

Any decent pantheon must have twelve gods. The ancient Greek gods had twelve: six men and six women. (I'd name them, but who cares?) Of course, they had demigods and other divinities, as well as superhuman mortals and a whole cast of interesting characters. But when it came to populating the Mount Olympus penthouse, twelve gods had the keys.

The traditional For Dummies book ends with The Part of Tens, in which each chapter contains ten items, such as tips, suggestions, things to remember, and other interesting tidbits. It's not that I mean to offend any two gods by limiting the lists in this part of the book to ten items. No, I respect the For Dummies tradition. So, in hindsight, it's probably a good thing that I didn't name all twelve gods in this introduction.

24

Ten Tips, Tricks, and Shortcuts

*I*f you look in the dictionary, you'll find several definitions for the word *tip*. The one I'm referring to for this chapter isn't the *tip* in the tip of your nose. Neither is it the *tip* that topples something, nor is it a gratuity. Nope, the tips in this chapter are handy, practical words of advice. Because not everyone considers good advice to be tip-worthy (where *tip* is a gratuity), I've also thrown in some tricks and shortcuts to even things out.

CopyService
Started by application: touch t

Google Voice
Process: com.google.android.apps.googl

WidgetService
Started by application: touch t

UpdateService
Started by application: touch t

Summon a Recently Opened App

I have to kick myself in the head every time I return to the All Apps screen to, once again, scroll to the depths to dig up an app I just opened. Why bother? Because I can press and hold the Home soft button to instantly see a list of recently opened apps.

Pressing and holding the Home soft button works no matter what you're doing on the ThunderBolt; you don't necessarily have to view the Home screen to see the list of recently opened apps.

Stop Unneeded Services

Some things may be going on in your ThunderBolt that you don't need or even suspect. These activities include the monitoring of information and the updating of apps, and tiny programs sometimes even check on the device's status. The technical term for these activities is *services*.

When a service has started that you don't want, or have been requested to stop, you can halt the service. Here's how:

1. **While at the Home screen, press the Menu soft button.**

2. **Choose Settings.**

3. **Choose Applications.**

4. **Choose Running Services.**

   You see the Running Services screen, similar to Figure 24-1.

5. **Touch a service to stop it.**

   Most likely, it's a service you recognize that you don't need or a service you've been directed to disable from another source or authority.

6. **Touch the Stop button to halt the service.**

When you stop a service, you free the resources used by that service. These resources include memory and processor power. The result of stopping unneeded services can be improved phone performance.

The service you stopped will most likely start up again the next time you start the ThunderBolt, or if you run the app. The only way to halt a specific service for all eternity is to uninstall the program associated with the service, which is a drastic step. Even then, not every app can be removed; preinstalled apps and phone company apps are stuck to your ThunderBolt like barnacles on a barge.

✔ Do not randomly disable services. Many of them are required for the ThunderBolt to do its job, or for the apps you use to carry out their tasks. If you disable a service that you don't recognize and the device begins to act funny, turn the ThunderBolt off and then on again. That should fix the problem.

✔ The colorful bar at the bottom of the Running Services screen (refer to Figure 24-1) illustrates memory usage in your ThunderBolt. The red area represents services that cannot be stopped or killed. In Figure 24-1, four apps are represented there, which occupy a total of 84MB of storage. The yellow area represents services used by 25 apps that occupy 232MB. The green area represents 144MB of free space.

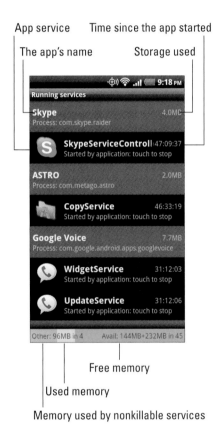

Figure 24-1: Services running on the ThunderBolt.

Set Keyboard Feedback

Typing on a touchscreen keyboard isn't easy. Along with the screen being tiny (or your fingers being big), it's difficult to tell what you're typing. You can add some feedback to the typing process. Heed these steps:

1. **While at the Home screen, press the Menu soft button.**
2. **Choose Settings.**
3. **Choose Sound Settings.**
4. **Put a check mark by the option Vibrate Feedback.**

   This option causes physical feedback when you press a "key" on the onscreen keyboard.

Now when you type on the onscreen keyboard, you can feel the keys as you press them.

The Vibrate Feedback setting also applies to the soft buttons and to other things you touch in the ThunderBolt interface.

Add Spice to Dictation

I feel that too few people use dictation, despite how handy it can be — especially for text messaging. Anyway, if you've used dictation, you might have noticed that it occasionally censors some of the words you utter. Perhaps you're the kind of person who doesn't put up with that kind of ####.

Relax, ####. You can lift the vocal censorship ban by following these steps:

1. **At the Home screen, press the Menu soft button.**
2. **Choose Settings.**
3. **Choose Voice Input and Output.**
4. **Choose Voice Recognizer Settings.**
5. **Remove the check mark by the option Block Offensive Words.**

And just what are offensive words? I would think that *censorship* would be an offensive word. But no, apparently the words ####, ####, and even innocent little old #### are deemed offensive by Google Voice. What the ####?

Add a Word to the Dictionary

Betcha didn't know that the ThunderBolt has a dictionary. It's used to keep track of words you type — words that may not be recognized as being spelled properly.

Words can be added to the dictionary as you type on the phone and the words come up. Refer to Chapter 4 to review the technique. To see the words in the dictionary, and manually add even more, follow these steps:

1. **On the Settings screen, choose Language & Keyboard.**

2. **Choose Touch Input.**

3. **Choose Personal Dictionary.**

4. **Choose Edit Personal Dictionary.**

   The User Dictionary screen appears. Listed are all the words you've added to the dictionary.

To add a new word, touch the Add New button and type the word; touch the OK button.

To remove a word from your personal dictionary, press the Menu soft button and choose the Delete command. Place a red X by every word you want to delete. Touch the Delete button to purge the selected words.

Create a Direct-Dial Screen Shortcut

For the numbers you dial most frequently, use the Favorites list, as described in Chapter 5. For your überfavorites, use Home screen direct-dial shortcuts. Here's how to create a direct-dial contact shortcut on the Home screen for someone you call frequently:

1. **Long-press a blank part of the Home screen.**

2. **Choose Shortcut from the Personalize screen.**

3. **Choose Direct Dial.**

4. **Choose the contact you want to direct-dial.**

5. **Choose the phone number, if the contact has multiple phones.**

   The direct-dial shortcut is placed on the Home screen you long-pressed in Step 1. If the contact has a picture, it appears as the shortcut's icon.

Touching the shortcut icon immediately dials the contact.

You can also create an icon to directly text-message a contact. The difference is that you choose Direct Message in Step 3.

Find Your Lost ThunderBolt

Someday, you may lose your ThunderBolt. It might be for a panic-filled few seconds — or for forever. The hardware solution is to weld a heavy object to the phone, such as an anvil or a school bus, yet that strategy kind of defeats the entire mobile/wireless paradigm. The software solution is to use a cell phone locator service.

A cell phone locator service employs apps that use the cellular signal as well as the phone's GPS to help locate a missing gizmo. These types of apps are available on the Android Market. I've not tried them all, and many of them require a subscription service or registration at a website to complete the process.

Here are some suggestions for cell phone locator apps:

- LocService
- Lookout Mobile Security
- Mobile Phone Locator

Enter Location Information for Your Events

When you create an event for the Calendar app, be sure to enter the event location. You can type either an address (if you know it) or the name of the location. The key is to type the text as you would type it in the Maps app when searching for a location. That way, you can touch the event location and the ThunderBolt displays it on the touchscreen. Finding an appointment couldn't be easier.

- See Chapter 13 for more information about the Maps app.
- See Chapter 17 for details about the Calendar.

Try the Quick Lookup App

I keep forgetting about the Quick Lookup app. It is found on the All Apps screen and occasionally accessed by way of other apps. You can use the app to search Google, Wikipedia, or YouTube, to translate text, or to refer to an online dictionary to define a term. It's a handy app, easily overlooked, yet quite powerful and useful.

In fact, to prevent overlooking the Quick Lookup app, consider placing a shortcut to it on the Home screen. See Chapter 22 for instructions.

Reset Your Google Password

Even though the computer security experts say you should change your passwords often, few people bother. Apparently, computer security experts carry less weight than the final eight contestants on *American Idol.* Despite that, when the mood hits you and you change your Google password, you need to update the ThunderBolt with the new password information. Follow these steps:

1. **On your computer, direct the web browser to** www.google.com **— the main Google page.**

2. **From the top of the page, click your account link.**

   As I write this chapter, the link is found in the upper right part of the page and shows your name as it's registered with Google.

3. **Choose Account Settings from the menu.**

4. **By the Security heading, click the link Changing Your Password.**

5. **Obey the directions on the screen for setting a new password.**

   For example, type your current password and then type your new password twice. Click the Save button.

   After your password has been reset, you need to update the ThunderBolt with the new password. If you don't, you see incessant error messages and the phone pesters you until you want to hurl it out a window. So continue with these steps:

6. **Wake up or turn on your phone.**

   In a few moments, you see the Alert icon appear at the top of the touchscreen display, in the notification area.

7. **Slide down the notification area by swiping it with your finger.**

   The specifics for performing this action are covered in Chapter 3.

8. **From the list of notifications, choose Alert.**

   The Alert message says Sign-In Error or Sign into Your Account.

9. **Type your new Google password in the box that appears on the touchscreen display.**

After you enter the new password, the phone changes its mood back to content and your Google account continues to sync.

Press the Home soft button to return to the Home screen.

Ten Things Worth Remembering

In This Chapter

▶ Locking the phone

▶ Saving typing time

▶ Speaking to the ThunderBolt

▶ Turning the phone sideways

▶ Minding the battery hogs

▶ Popping out the kickstand

▶ Watching out for a roaming signal

▶ Mounting the MicroSD card

▶ Taking a picture of a contact

▶ Using the Search command

*I*t's easier to come up with a list of things to remember than it is to come up with a list of things to forget. And I'm not talking about things to forget, as in *don't forget,* which is merely the same thing as *remember.*

For example, you should forget to use your ThunderBolt as an ice pick. Forget about painting the phone. And please forget that the phone makes a lousy comb. Instead, you can concentrate on remembering the ten items listed in this chapter, which are worthy of storage in your brain's memory.

. email...

‹o: Barack Obama potus@whit

message:

Lock the Phone on a Call

Whether you dialed out or someone dialed in, after you start talk-ing, you should lock your phone: Press the Power Lock button atop the ThunderBolt. By doing so, you ensure that the touchscreen is disabled and the call isn't unintentionally disconnected.

Of course, the call can still be disconnected by a dropped signal or by the other party getting all huffy and hanging up on you, but by locking the phone, you prevent a stray finger or your pocket from disconnecting (or muting) the phone.

Use the Keyboard Suggestions

Don't forget to take advantage of the suggestions that appear above the onscreen keyboard when you're typing text. In fact, you don't even need to touch a suggestion; to replace your text with the highlighted suggestion, simply touch the onscreen keyboard's Space key. Zap! The highlighted word appears.

If the keyboard suggestions don't show up, follow these steps to ensure that the predictive text feature has been activated:

1. **Start the Settings app.**

2. **Choose Language & Keyboard.**

3. **Choose Touch Input.**

4. **Choose Text Input.**

5. **Ensure that there's a check mark by the Prediction item.**

Also refer to Chapter 4 for additional information on using the keyboard suggestions.

Dictate Your Text

It's such a handy feature, yet I constantly forget to use it: Rather than type short text messages, use dictation. You can access dictation from any onscreen keyboard by touching the Microphone button. Speak the text, the text appears. Simple.

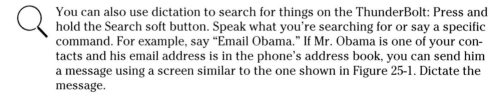

You can also use dictation to search for things on the ThunderBolt: Press and hold the Search soft button. Speak what you're searching for or say a specific command. For example, say "Email Obama." If Mr. Obama is one of your contacts and his email address is in the phone's address book, you can send him a message using a screen similar to the one shown in Figure 25-1. Dictate the message.

Figure 25-1: Dictating a quick email.

You can utter many other commands using dictation on the ThunderBolt. Just touch the Help button after summoning the main dictation screen.

Landscape Orientation

I enjoy the occasional widescreen view. Apps such as the Internet, Kindle, and even Mail look much better longways than "tall-ways." Tilting the ThunderBolt to the side can even help your typing, by presenting a wider version of the onscreen keyboard.

Not every app supports landscape orientation. Some apps, especially games, appear only in landscape orientation.

Things That Consume Lots of Battery Juice

Four items on the ThunderBolt suck down battery power faster than the massive alien fleet that's defeated by a plucky antihero who just wants the girl:

- ✔ The 4G LTE cellular data signal
- ✔ Navigation
- ✔ Bluetooth
- ✔ Wi-Fi networking

You probably bought the ThunderBolt because of its 4G LTE service. I don't blame you; it's terrifically fast! But the zippy wireless signal sucks up battery power like Congress spends money. Enjoy the speed, but keep an eye on the battery meter. And review Chapter 23 for battery-saving tips.

Navigation is certainly handy, but because the phone's touchscreen is on the entire time and dictating text to you, the battery drains rapidly. If possible, try to plug the phone into the car's power socket when you're navigating. If you can't, keep an eye on the battery meter.

Both Bluetooth and Wi-Fi networking require extra power for their wireless radios. When you need that speed or connectivity, they're great! I try to plug my phone into a power source when I'm accessing Wi-Fi or using Bluetooth. Otherwise, I disconnect from those networks as soon as I'm done, to save power.

- ✔ Technically speaking, using Wi-Fi doesn't drain the battery as drastically as you might think. In fact, the Wi-Fi signal times itself out after about 15 minutes of non-use. So it's perfectly okay to leave Wi-Fi on all day and you'll experience only a modicum of battery loss because of it. Even so, I'm a stickler for turning off the Wi-Fi when I don't use it.
- ✔ See Chapter 23 for more information on managing the phone's battery.

Use the Kickstand

I tend to keep my ThunderBolt in one spot when I'm not on the road. I just pop out its kickstand and prop up the phone in vertical orientation. The USB cable connects to the phone's lower left, which means that the thing is stable on my desktop.

You can also get a docking stand for the ThunderBolt. None is available to me as this book goes to press; otherwise, I'd writing glowing statements about them. (Keith, if you're reading this, you should have sent me a docking station.)

Check for Roaming

Roaming can be expensive. The last non-smartphone (dumbphone?) I owned racked up $180 in roaming charges the month before I switched to a better cellular plan. Even though you too may have a good cell plan, keep an eye on the phone's status bar. Ensure that when you're making a call, you don't see the Roaming status icon on the status bar atop the touchscreen.

Well, yes, it's okay to make a call when your phone is roaming. My advice is to remember to *check* for the icon, not to avoid it. If possible, try to make your phone calls when you're back in your cellular service's coverage area. If you can't, make the phone call but keep in mind that you will be charged roaming fees. They ain't cheap.

Properly Access the MicroSD Card

The ThunderBolt's removable storage area, the *MicroSD card,* can be accessed from your computer. To let you do so, the card needs to be mounted into the computer's storage system. That way, all the stuff you have on your ThunderBolt — music, videos, still pictures, contacts, and other types of information — can be accessed or backed up on the computer. Likewise, you can copy information from your computer to the MicroSD card on the ThunderBolt.

Oh, the whole accessing-the-MicroSD-card thing is an old song-and-dance that I go into detail about in Chapter 20. What's important to remember is that when the MicroSD card is mounted on your computer's storage system, the ThunderBolt cannot access the card. If an attempt is made, a message appears, explaining that the MicroSD card is busy.

When you're done accessing the MicroSD card from your computer, be sure to stop USB storage: Choose the Charge Only USB connection as a part of properly unmounting the phone from the computer's storage system. Refer to Chapter 20 for details.

Snap a Pic of That Contact

Here's something I always forget: Whenever you're near one of your contacts, take the person's picture. Sure, some people are bashful, but most folks are flattered. The idea is to build up the People app's address book so that all contacts have photos.

When taking a picture, be sure to show it to the person before you assign it to the contact. Let them decide whether it's good enough. Or, if you just want to be rude, assign a crummy-looking picture. Heck, you don't even have to do that: Just take a random picture of anything and assign it to a contact: A plant. A rock. Your cat. But, seriously, keep in mind that the phone can take a contact's picture the next time you meet up with that person.

See Chapter 15 for more information on using the ThunderBolt's camera and assigning a picture to a contact.

The Search Command

Google is known worldwide for its searching abilities. By gum, the word *Google* is now synonymous with searching. So please don't forget that the ThunderBolt, which uses the Google Android operating system, has a powerful Search command.

The Search command is not only powerful but also available all over. The Search soft button can be pressed at any time, in just about any program, to search for information, locations, people — you name it. It's handy. It's everywhere. Use it.

Ten Nifty Apps

C hapter 18 shows that you can fill your ThunderBolt with as many apps as will fit inside. Considering that the average app occupies about .003 potrzebies of storage, that makes for a lot of apps you can potentially store in the phone. But which apps?

To help you get started on your app-collection journey, I offer you this chapter. In it, I present you with ten apps I recommend, apps that will help you start your app library. All apps listed here are free, and they do things that the preinstalled apps on the ThunderBolt don't do, or don't do well. Enjoy!

.izon Wireless

Google Books   Update (r
Google Inc.   ☆

Dropbox   Update (r
Dropbox, Inc.   ☆

Angry Birds Seasons
Rovio Mobile Ltd.   ☆

ⁿkick

Angry Birds

 The birds may be angry at the green piggies for stealing eggs, but you'll be crazy for this addictive game. Like most popular games, Angry Birds is simple. It's easy to learn, fun to play.

You can find a series of Angry Birds games, including one for holidays and another for an adventure the birds have in Rio. So there's plenty of bird action if you get addicted to this game.

Barcode Scanner

Many apps from the Android Market can be quickly accessed by scanning their barcode information. Scanning with what? Why, your ThunderBolt, of course!

By using an app such as Barcode Scanner, you can instantly read in and translate bar codes that open product descriptions, web page links, or links directly to apps in the Android Market.

Though you can find similar barcode scanning apps, I find Barcode Scanner the easiest to use: Run the app. Point the phone's camera at a bar code and, in a few moments, you see a link or an option indicating what to do next. To get an app, choose Open Browser. This option takes you to the Android Market, where obtaining the app is just a few touches away.

CardStar

The handy CardStar app answers the question "Why do I have all these store cards?" They're not credit cards — they're marketing cards designed for customer rewards or loyalty programs. Rather than tote those cards around in your wallet or on your keychain, you can scan a card's bar code using your ThunderBolt and save the "card" on the phone.

After you store your loyalty cards in the ThunderBolt, you simply run the CardStar app to summon the appropriate merchant. Show the checkout person your phone or scan the bar code yourself. CardStar makes it easy.

Dolphin Browser

The Internet app that comes with the ThunderBolt is despised by many Android users. A better and more popular alternative is Dolphin Browser.

Like many popular computer browsers, Dolphin Browser features a tabbed interface, which works much better than the silly multiple-window interface of the boring Internet app.

Using Barcode Scanner in this chapter

Throughout this chapter, you find the barcode icon, or, specifically, the *QR code,* which is a special square type of bar code. After installing the Barcode Scanner app (or a similar app), use your ThunderBolt to scan a bar code. Touch the Open Browser button to download its recommended app from the Android Market.

Additional app recommendations and their bar codes are found all over the Internet, in magazines, and on my ThunderBolt support page at www.wambooli.com/help/phone.

Dolphin Browser also sports many handy tools, which you can access by pressing the Menu soft button. Unlike on other Android apps, the tools pop up on a menu that you can see on the screen.

Gesture Search

The Gesture Search app provides a new way to find information on your ThunderBolt. Rather than use a keyboard or dictate, you simply draw on the touchscreen the first letter of whatever you're searching for.

Start the Google Search app to begin a search. Use your finger to draw a big letter on the screen. After you draw a letter, search results appear on the screen. You can continue drawing more letters to refine the search or touch a search result.

Gesture Search can find contacts, music, apps, and bookmarks in the Internet app.

Google Finance

The Google Finance app is an excellent market-tracking tool for folks who are obsessed with the stock market or want to keep an eye on their portfolios. The app offers you an overview of the market and updates to your stocks as well as links to financial news.

To get the most from this app, configure Google Finance on the web, using your computer. You can create lists of stocks to watch, which are then instantly synchronized with your ThunderBolt. You can visit Google Finance on the web at

```
www.google.com/finance
```

As with other Google services, Google Finance is provided to you for free, as part of your Google account.

Google Sky Map

 Ever look up at the sky and say, "What the heck is that?" Unless it's a bird, an airplane, a satellite, or a UFO, Google Sky Map will help you find what it is. You may learn that a particularly bright star in the sky is, in fact, the planet Jupiter.

The Google Sky Map app is elegant. It basically turns the ThunderBolt into a window you can look through to identify objects in the night sky. Just start the app and hold the ThunderBolt up to the sky. Pan the phone to identify planets, stars, and constellations.

 Google Sky Map promotes using the ThunderBolt without touching it. For this reason, the screen goes blank after a spell, which is merely the phone's power-saving mode. If you plan extensive stargazing using Google Sky Map, consider resetting the screen timeout. Refer to Chapter 2 for details.

Movies

 The Movies app is the ThunderBolt's gateway to Hollywood. The app lists currently running films and films that are opening, and it has links to your local theaters with showtimes and other information. The app is also tied into the popular Rotten Tomatoes website for reviews and feedback. If you enjoy going to the movies, you'll find the Movies app a valuable addition to your ThunderBolt's app library.

SportsTap

 I admit to not being a sports nut, so it's difficult for me to identify with the craving to have the latest scores, news, and schedules. The sports nuts in my life, however, tell me that the very best app for this purpose is a handy thing named SportsTap.

Rather than blather on about something I'm not into, just take my advice and obtain SportsTap. I believe you'll be thrilled.

Zedge

 The Zedge program is a helpful resource for finding wallpapers and ringtones — millions of them. It's a sharing app, so you can access wallpapers and ringtones created by other Android users as well as share your own. If you're looking for a specific sound, or something special for Home screen wallpaper, Zedge is the best place to start your search.

Index

• X •

• Y •

• Z •

Notes

Notes

4/12   **WITHDRAWAL**   2011

4c-7/17 (12/12)